Methods for Imaging Cell Membranes

This book will serve as an introduction to microscopy and biomedical imaging methods, with a focus on the study of the distributions and dynamics of molecules on the cell surface. It will provide readers with an in-depth understanding of how modern microscopy methodology can be used to understand the organisation of cell membrane systems and how experiments can be designed around these methodologies.

There are numerous methods employed to understand cell membrane organisation, but foremost among them are microscopy methods which can map the distributions of molecules on the cell surface and even map the biophysical properties of membranes themselves. Fluorescence microscopy has been especially widely used due to its specificity and relatively non-invasive nature, allowing live-cell imaging. However, the recent advance of super-resolution fluorescence microscopy has broken the previous resolution limit for this type of microscopy, which has been an important advancement in the field. Atomic force microscopy and electron microscopy have also been deployed to learn about membrane organisation and properties.

Each chapter in this volume will be themed around measuring a particular property of cell membranes. In each case, the authors examine the range of methodology applicable to the task, comparing the advantages and disadvantages of each one, and will also provide an overview of important discoveries that have been made using the methodology being discussed. The chapters will cover:

- Measuring membrane protein distributions using single-molecule localisation microscopy (SMLM).

- Measuring membrane protein dynamics and diffusion using fluorescence correlation spectroscopy (FCS).

- Mapping membrane lipid backing using environmentally sensitive fluorescence probes.

- Mapping membrane thickness and rigidity using atomic force microscopy.

- Mapping membrane proteins and the cytoskeleton using electron microscopy.

This book will be a valuable resource to graduate and upper-level undergraduate students and industry researchers in the fields of cell biology, microbiology, microscopy, and medical imaging.

SERIES IN CELLULAR AND CLINICAL IMAGING
Series Editor Ammasi Periasamy

The purpose of the series is to promote education and new research using cellular and clinical imaging techniques across a broad spectrum of disciplines. The series emphasizes practical aspects, with each volume focusing on a particular theme that may cross various imaging modalities. Each title covers basic to advanced imaging techniques as well as detailed discussion dealing with interpretations of the studies. The series provides cohesive, complete, and state-of-the-art cross-modality overviews of the most important and timely areas within cellular and clinical imaging.

For more information about this series, please visit: [https://www.crcpress.com/Series-in-Cellular-and-Clinical-Imaging/book-series/CRCSERCELCLI]

Methods for Imaging Cell Membranes

Luca Panconi, Daniel Nieves,
Maria Makarova, and Dylan Owen

CRC Press
Taylor & Francis Group
Boca Raton London New York

CRC Press is an imprint of the
Taylor & Francis Group, an **informa** business

Designed cover image: Shutterstock_465563519

First edition published 2024
by CRC Press
2385 NW Executive Center Drive, Suite 320, Boca Raton FL 33431

and by CRC Press
4 Park Square, Milton Park, Abingdon, Oxon, OX14 4RN

CRC Press is an imprint of Taylor & Francis Group, LLC

ISBN: 9781032207902 (hbk)
ISBN: 9781032226941(pbk)
ISBN: 9781003273745 (ebk)

DOI: 10.1201/9781003273745

Typeset in Minion
by codeMantra

Contents

Authors

Luca Panconi is a mathematician and bioinformatician who specialises in implementing analysis techniques for super-resolution microscopy data. His work crosses the boundaries of computer science and biology, creating innovative data analysis tools that unlock unprecedented details at a molecular level. As a statistician with a commitment to advancing the scientific community, his work bridges the gap between complex mathematics and biological research.

Dr. Daniel Nieves has been a research fellow in the lab of Prof. Dylan Owen within the Institute of Immunology and Immunotherapy at the University of Birmingham, UK, since 2019. His work has focused on implementation and design of a framework to aid cluster analysis implementation for single-molecule localisation microscopy. His research focuses on the development of new labelling and analysis methods to improve quantitative biological measurements using super-resolution microscopy.

Dr. Maria Makarova was promoted to Assistant Professor in 2022 in the School of Biosciences. Her work combines comparative microbiology, genetics, lipidomics, and advanced microscopy to understand the interplay of metabolism, the membrane, and the environment.

Prof. Dylan Owen moved to the University of Birmingham in 2019 and now holds an interdisciplinary chair position between the Institute of Immunology and Immunotherapy and the School of Mathematics, as well as serving as the deputy-director of the Centre of Membrane Proteins and Receptors (COMPARE). His lab's work seeks to develop a new microscopy methodology (especially single-molecule localisation microscopy, image analysis, and AI) and apply these methods to study membrane biophysics and T cell immunology.

CHAPTER 1

General Introduction to Imaging the Plasma Membrane

1.1 THE PLASMA MEMBRANE

1.1.1 General Structure

One of the most prevalent classes of biological molecules in any plasma membrane is the phospholipid, which contains a hydrophilic head connected to two acyl lipid (fatty acid) tails by a glycerol backbone.

The head group consists of a variable group, four to five of which are common (for example, choline, ethanolamine, serine, and inositol), linked to glycerol via a phosphate group. The fatty acids are predominantly comprised of a hydrocarbon chain, constructed from an even number of carbon atoms and typically range in length from 12 to 24 carbons, with 16 and 18 usually being the most prevalent (Gennis 1989). The fatty acid can be either saturated, where each carbon atom in the chain is bound to two hydrogen atoms and contains only single carbon–carbon bonds, or unsaturated, where some carbon atoms form double bonds with adjacent carbon atoms. In mammals, the fatty acids are linked to the glycerol backbone by ester bonds. Note that the structure described here only captures the major class of lipid found in mammalian cell membranes. Other classes include the sphingolipids which have a different backbone. In other domains of life,

DOI: 10.1201/9781003273745-1

lipids can be very different, for example, the absence of ester bonds or the presence of branched acyl chains to name but a few variations.

The membrane itself is composed of two layers, primarily composed of phospholipids, known as leaflets, with the hydrophilic heads facing outwards from the membrane, towards the aqueous surrounding medium, and hydrophobic acyl lipid tails directed inwards. This structure is appropriately defined as a phospholipid bilayer (Figure 1.1), which acts as the stable barrier separating the cell interior from the cell exterior and provides the underlying foundation of the membrane. Within this bilayer, other molecules are embedded. These include other lipids such as sterols which serve to regulate the membrane's biochemical and biophysical properties. Also present are membrane proteins, either targeted to the membrane by post-translational modification or via transmembrane domains. This configuration of a lipid bilayer and membrane-spanning, mobile proteins is known as the Fluid Mosaic Model or the Singer–Nicolson model (Singer and Nicolson 1972). In general, the two leaflets are not identical and will have different lipid compositions and biophysical properties (Rothman and Lenard 1977; Lorent, Leventhal et al. 2020).

Phospholipids with saturated acyl tails are highly linear and pack efficiently against neighbouring phospholipids. On the contrary, unsaturated

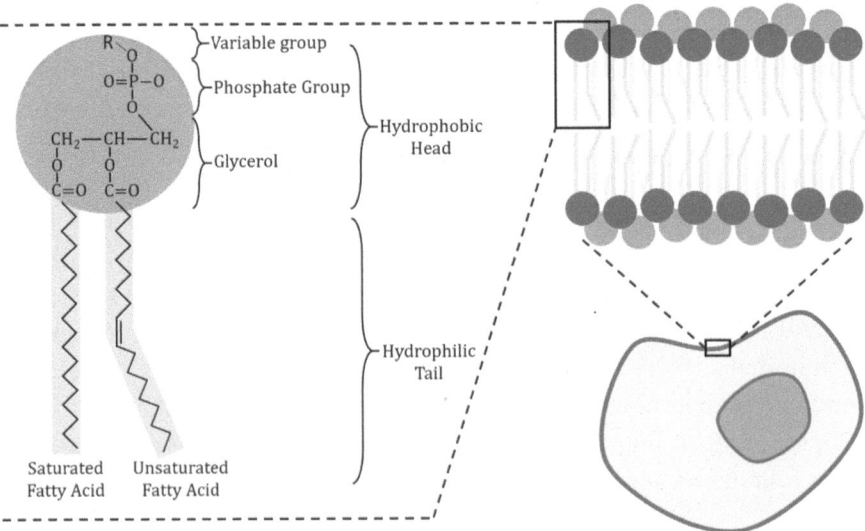

FIGURE 1.1 Molecular structure of an individual phospholipid and its position within the phospholipid bilayer of a plasma membrane.

lipid tails can adopt trans or cis configurations at the double-bond position and become kinked, reducing the overall packing efficiency and reducing the lipid's melting temperature. If more than one double bond is present, they will occur at three carbon intervals – this is known as the divinylmethane pattern. Due to the differences in melting temperatures between saturated and unsaturated lipids, a phospholipid bilayer can exist in either a liquid or gel state depending on the phospholipid composition and temperature. These states are called phases and are mainly dictated by the structure of the lipid tails (Figure 1.2). Membranes formed from a combination of saturated and unsaturated lipids can phase separate with liquid and gel regions existing in the same bilayer. Simple membranes such as these can readily be constructed in vitro from synthesised or extracted, purified lipids (Mouritsen and Bagatolli 2015).

1.1.2 Composition

Alongside phospholipids, the plasma membrane contains a diverse array of biological molecules. The presence and quantities of these molecules typically vary across species and even among different cells of the same organism. For example, mammalian cell plasma membranes contain glycolipids – molecules comprised of a carbohydrate molecule (in particular, a mono- or oligosaccharide group) attached to the phospholipids. Their abundance in the plasma membrane is highly variable between organisms and cell types, but they are located exclusively within the outer leaflet, and account for around 2% of lipids in most plasma membranes.

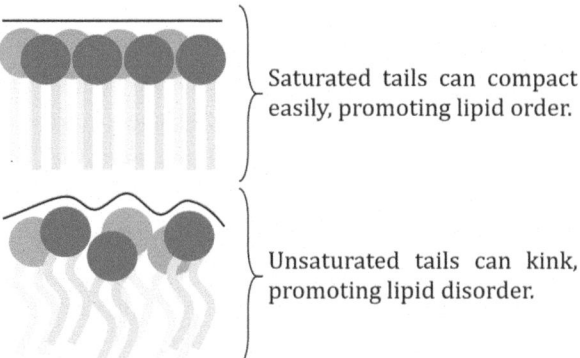

Saturated tails can compact easily, promoting lipid order.

Unsaturated tails can kink, promoting lipid disorder.

FIGURE 1.2 Differences between saturated and unsaturated lipids and their impact on lipid order. Saturated fatty acids compact easily, promoting lipid order, while unsaturated fatty acids contain kinks, which promote lipid disorder.

These molecules typically act to mediate the interactions of the cell with its environment, such as the extracellular matrix or other cells (Schnaar 2004).

Another class of molecules present in cell membranes are the sterols. In mammalian plasma membranes, cholesterol is the primary sterol species. Notably, cholesterol is highly prevalent in the plasma membranes of animal cells, often matching the molar quantity of the phospholipids (around 30% of all molecules). Cholesterol is a sterol (steroid alcohol), a 27-carbon compound comprising a hydrocarbon tail, four hydrocarbon rings (the sterol nucleus), and a hydroxyl group. The ring structure of the sterol nucleus is flat and rigid, allowing the molecule to fit within the spaces between phospholipids. Because the hydroxyl group is small and is the only polar part of the molecule, cholesterol is not able to form bilayers by itself. Instead, it "shelters" under the phospholipid head groups, among the lipid acyl tails.

The role of cholesterol is to regulate membrane fluidity, although the exact effect depends on temperature. At high temperatures (when the bilayer would naturally be in the liquid phase), cholesterol imposes order of adjacent phospholipid tails, therefore yielding greater lipid packing. This induced a new membrane phase – the liquid-ordered phase, meaning the previous (non-sterol) phase is usually termed the liquid-disordered phase. At low temperatures, the opposite is true: cholesterol interferes with fatty acid chain packing, maintaining membrane fluidity and reducing the risk of freezing. Overall, the liquid-ordered phase combines the desirable properties of the liquid-disordered and the gel phases. Component molecules remain laterally mobile within the plane of the bilayer (like the liquid-disordered phase), but the bilayer has higher thickness, rigidity, and resistance to solubilisation. Like the liquid and gel phases, membranes composed of cholesterol can phase separate into co-existing liquid-disordered and liquid-ordered regions. Typically, the liquid-ordered regions are termed lipid rafts or membrane lipid microdomains. The existence of this liquid–liquid phase separation in cell membranes is termed the Lipid Raft Hypothesis (Simons and Ikonen 1997) and has proven controversial (Munro 2003).

Proteins, one of the most important types of biological molecules associated with the cell, are ubiquitous in the plasma membranes of all life forms. Proteins associated with the bilayer may be integral (embedded within the membrane) or peripheral (attached to the surface via protein–protein interactions, usually involving ionic bonds) and are responsible for regulating cellular processes and membrane function. Most integral proteins are transmembrane proteins, meaning part of their structure is exposed on both the exterior and interior leaflets. These transmembrane proteins are fixed in the membrane by the hydrophobic amino acids that are present within the polypeptide chain. Carbohydrates are often added during the construction of these polypeptide chains, and so most transmembrane proteins are glycoproteins. Some proteins, such as porins, form channels which selectively allow molecules to pass into the cell. Plasma membranes are composed of roughly equal masses of lipids and proteins; however, protein molecules are typically much denser than lipids, and so they only comprise 1%–2% of membrane molecules by number (Figure 1.3). Because of this, the plasma membrane has been termed a "lipid–protein composite" (Jacobson, Mouritsen et al. 2007).

Broadly speaking, there are four different varieties of integral membrane proteins, as follows:

- Linkers and adhesion molecules: These support cell structure by anchoring the cortical actin cytoskeleton to the membrane and to the extracellular matrix as well as adhering to extracellular substrates or to other cells. Integrins are a common example of this type of membrane protein (Hynes 2002).

- Enzymes: These regulate metabolism and signalling in the vicinity of the membrane. A large fraction of metabolic reactions occur in close

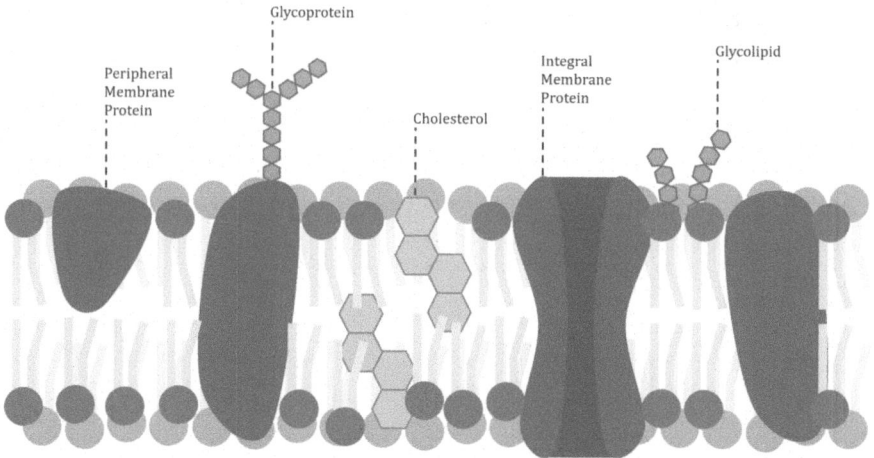

FIGURE 1.3 The primary components of the plasma membrane, including the lipid bilayer, membrane proteins, glycoproteins, glycolipids, and cholesterol. This is representative of the fluid mosaic model.

proximity to membranes. Enzymes are frequently utilised in signal transduction and are often post-translationally modified to target them to membranes. The Ras family of GTPases, commonly found mutated in cancers, are a good example of acylated, membrane-targeted enzymes (Hancock, Magee et al. 1989).

- Pumps and channels: These are responsible for the transport of ions, sugars, and amino acids and other molecules across the membrane. Channels are passive, allowing diffusion from high to low concentration, whereas pumps consume energy, usually in the form of ATP, and can transport molecules against a concentration gradient. For example, the sodium–potassium ATPase ion pump is important for maintaining membrane voltage, and neurons expend a considerable fraction of their energy budget on its operation (Clausen, Hilbers et al. 2017).

- Receptors: These are binding sites for ligands that initialise signalling pathways. These allow the cell to sense and respond to its extracellular environment and communicate with other cells. The family of G-protein-coupled receptors (GPCRs) are a major class of receptors and are frequent drug targets in the pharmaceutical industry (Lindsley 2013).

Proteins (like lipids) can show varying affinity for the different possible bilayer phases. For example, proteins with longer transmembrane domains, or proteins post-translationally modified with long, saturated acyl tails will typically show affinity of the liquid-ordered phase. This bilayer phase separation can be a mechanism for the regulation of protein–protein interactions at the membrane as well as a way to modulate membrane mechanical properties (Lingwood and Simons 2010; Owen, Williamson et al. 2012).

1.1.3 Function

Generally, the plasma membrane acts as the barrier between the internal aqueous compartment comprising the cell interior and the surrounding medium. Its most basic, primary function is to keep the organelles, molecules, and cellular features associated with the cell isolated from other biological structures. However, it also acts as the foundation for complex biological processes. As discussed, membrane proteins are responsible for carrying out the more specialised functions of the membrane. These include cell communication, cell–cell recognition, and active transport of molecules from the surrounding medium. However, since the interactions of membrane proteins are dependent on the underlying membrane phase, the diversity of phospholipids also contributes to the efficiency of these additional functions. For instance, the phospholipid known as phosphatidylinositol, although a relatively minor membrane component, is essential for cell signalling (Michell 1975). Some examples of the functions of the plasma membrane are summarised below:

- **Cell signalling:** An umbrella term for any system by which the plasma membrane receives and processes information from the cell's environment. This is typically achieved through cell surface receptors – these are proteins or glycoproteins which have highly specific binding sites for particular ligands. This binding will usually bring about activation of enzymes associated with the membrane, altering cell metabolism and ultimately giving rise to a cascading chemical change. GPCRs, also known as 7-transmembrane domain receptors, are one of the largest classes of membrane receptors. Immune receptors (such as the TCR or BCR), hormone receptors, and adhesion receptors (such as the integrins mentioned earlier) are other examples. Ligands for membrane receptors might be soluble or presented on the surface of other cells (Figure 1.4).

- **Selective transport:** The extracellular medium surrounding each cell contains nutrients that the cell requires for survival and growth.

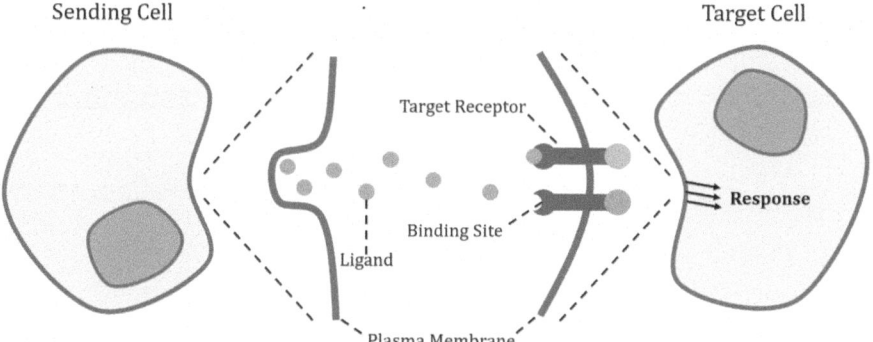

FIGURE 1.4 The process of cell signalling. The sending cell releases ligands which bind to target receptors embedded in the plasma membrane of a target cell. This brings about a response within the target cell.

However, the cell environment will typically also contain a number of other harmful or toxic substances. As such, the bilayer of the plasma membrane restricts access from most biological molecules, but is populated with a range of specialised transmembrane proteins that permit specific molecules and ions. Some proteins form channels which only open in the presence of particular molecules (ligand-gated channels), and these are known as facilitators or channel proteins. Other proteins, known as pumps, force solutes through the membrane. In general, channels and pumps can be selective or more general, and they can be gated by ligands, voltage, mechanical force, or other means.

- **Endocytosis and exocytosis:** The process by which a region of the extracellular medium is internalised. Small membrane protrusions wrap around the target region, with the target molecules contained therein, and engulf the medium. This is then transported into the cell in a small membrane compartment known as a vesicle. There are various types of endocytosis depending on the mechanism of regulation, such as receptor-mediated endocytosis. Exocytosis is the antithesis to endocytosis, in which vesicles of molecules to be externalised from the cell are brought to the plasma membrane. The vesicle is absorbed into the membrane and in doing so expels its contents into the extracellular medium. The release of neurotransmitter at the neuronal synapse is a classic example of exocytosis (Figure 1.5).

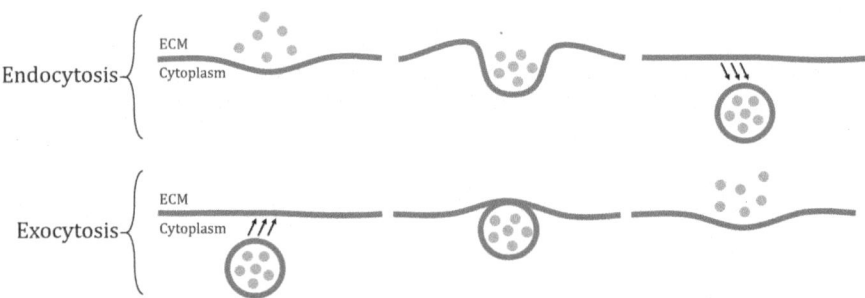

FIGURE 1.5 The stages of endocytosis and exocytosis. During endocytosis, large particles are ingested by the membrane and transferred into the cell cytoplasm in the form of a vesicle. During exocytosis, vesicles are brought to the plasma membrane. The vesicles fuse with the membrane, and their contents are released into the extracellular environment.

1.1.4 Membrane Diversity

Despite differences in molecular composition, all biological membranes follow the fluid mosaic model, but with the presence or absence of particular membrane constituents. For instance, membranous organelles, such as the rough endoplasmic reticulum, mitochondria, and the nucleus, are specialised structures, encompassed by their own membrane, which carry out particular functions within cells. Each of these organelles presents membranes that differ from the plasma membrane in composition and organisation. In general, intracellular membranes such as these contain lower concentrations of sterols and higher fractions of lipids with unsaturated acyl tails. As such, they have a higher abundance of the liquid-disordered phase, with intracellular membranes consequently more fluid than the plasma membrane (Owen, Rentero et al. 2012).

As discussed, membranes can differ in structure and composition on an intracellular basis. However, they also vary between different cell types, even in the same multicellular organism. For example, ciliated epithelial cells in humans possess long thread-like protrusions called cilia. These are composed of microtubules that are coated in the cell plasma membrane and are responsible for locomotion of fluids, such as mucus, on the cell surface. Immune cell membranes, such as the T (thymus) cells, contain specific receptors for identifying antigens – molecules that are associated with pathogens – and therefore assist in the immune response. A neuronal cell has a specialised plasma membrane, which is highly polarised, exhibits large membrane protrusions, and contains an excess of sodium, potassium, and chlorine ion channels. All of these features assist the neuron in achieving conduction of electrochemical signals across the nervous system. As is the case with membranous organelles, the membrane of each cell is adapted to carry out its specific cell function.

The existence and quantities of biological molecules within the plasma membrane can differ dramatically across species. For example, cholesterol is extremely prevalent in mammalian cell plasma membranes, but it is absent from bacterial and plant cell membranes. Instead, these species utilise different sterols to regulate membrane fluidity. Some species will also display entirely different structures attached to their membranes. Bacteria, such as *Escherichia coli*, express a dual-membrane system, composed of two separate plasma membranes, in which the inner membrane is surrounded by both a periplasmic space and an additional outer membrane. The outer membrane differs from traditional plasma membranes as it is highly permeable to ions and small polar molecules. This is on account of the increased prevalence of porins, which form open aqueous channels through the membrane. Some of these membrane proteins are shared with the membranes of mitochondria and chloroplasts in plant cells. This is, in part, evidence of the endosymbiotic hypothesis that mitochondria and chloroplasts are descendants of ancient, specialised bacteria that survived endocytosis by other cell types. Interestingly, some studies have suggested the evolutionary convergence of membrane properties between species (Kaiser, Surma et al. 2011).

1.2 INTRODUCTION TO FLUORESCENCE IMAGING

In most cases, cells range in size from around a micron to 100 μm. So, in order to investigate the structure, composition, and function of the plasma membrane, advanced microscopy systems must be implemented. One such microscopy system is fluorescence microscopy, which exploits the principles of fluorescence to identify cellular structures. This section explores the properties of fluorescent molecules and highlights how they can be used to achieve cellular imaging at high spatial and temporal resolutions.

1.2.1 What Is Fluorescence?

Fluorescence is defined as the process of absorbing light at one particular wavelength and then re-radiating light at a different (longer) wavelength. From a physicochemical perspective, this corresponds to the excitation of an electron to a higher energy state, loss of some energy to the surrounding medium, and return of the electron to the ground state resulting in the emission of light with a lower energy, longer wavelength. The entire fluorescence process is transient, typically lasting several nanoseconds (the fluorescence lifetime – the time the electron spends in the excited state). Chemical compounds that display properties conducive to fluorescence are known as fluorophores. An example of the chemical structure of a commonly used fluorophore – Alexa Fluor 488 – is shown in Figure 1.6.

In order to conduct fluorescence microscopy, samples must first be stained with fluorophores such as Alexa 488. Thankfully, a range of different fluorophores exist for staining different cellular components, making it possible to identify these components with a high degree of specificity, and the plasma membrane is no exception. There are membrane dyes, which mimic phospholipids of sterols in structure and when added to cells will incorporate into membranes. It is also possible to label specific membrane proteins, e.g., through the expression of fluorescent fusion constructs or

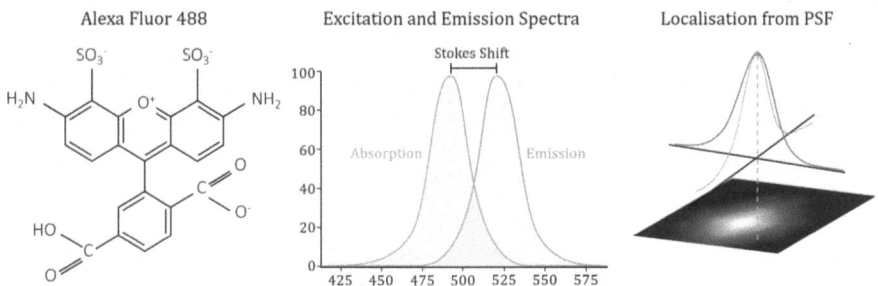

FIGURE 1.6 The structure and properties of the fluorophore Alexa Fluor 488, including chemical structure, emission and excitation spectra, and typical point spread function that would be observed on the camera.

through immunostaining using antibodies conjugated to small molecule organic fluorophores.

The range of wavelengths of light emitted by a fluorophore is known as its emission spectrum. This typically depicts the relative intensity of each wavelength of light which is emitted by the fluorophore. Each fluorophore also has a specific range of excitation wavelengths, known as the excitation spectrum, which will have a range of wavelengths that are shorter than that of the emission spectrum due to the loss of energy during electron relaxation to the bottom of the excited state. This difference in peaks between the emission and excitation spectra is known as the Stokes shift (Figure 1.6). Ideally, the Stokes shift will be as large as possible to distinguish light emitted from the sample and light used for excitation. The length of time that an electron spends in the excited state between excitation and emission is known as the fluorescence lifetime. This is typically several nanoseconds and depends on both the fluorophore and its local environment. Measuring this time, as done in fluorescence lifetime imaging, can therefore be used to probe a fluorophore's surroundings.

1.2.2 The Basic Fluorescence Microscope

There are a myriad of different designs of fluorescence microscope, each with pros and cons and chosen based on the application (Yuste 2005). A fluorescence microscope (Figure 1.7) requires a powerful light source to illuminate the sample. Typically, lasers are used as these are powerful, monochromatic, and can be focused to a small spot. LEDs or arc lamps can also be used. In common epi-fluorescence system configurations, excitation light is directed to the sample via a dichroic mirror and then through the microscope objective lens. This configuration is particularly

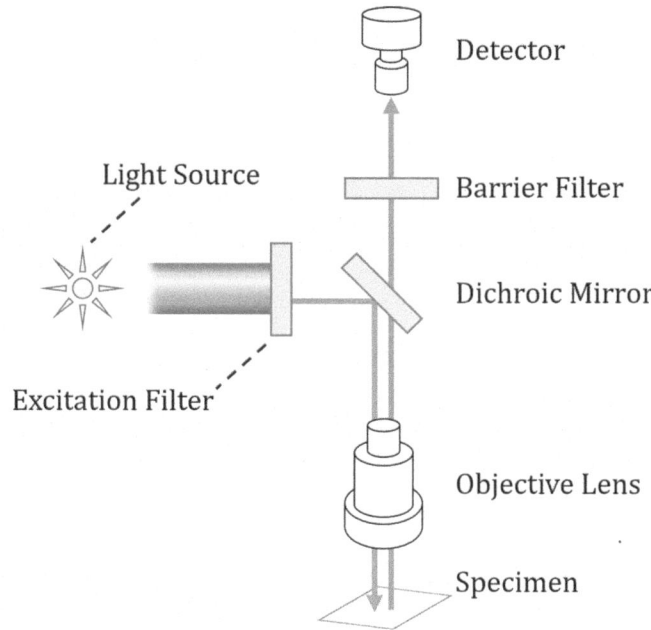

FIGURE 1.7   Optical diagram of a fluorescence microscope, with light source filtered through a dichroic mirror, an objective lens, and onto a sample. Fluorescence light is filtered back through a barrier filter and onto the detector.

useful when studying live cells because an inverted microscope can be used with the objective placed underneath the sample, allowing cells to be imaged in an aqueous environment, such as cell culture media.

Light emitted from fluorescent molecules is collected back through the objective lens and is then filtered back through the dichroic and a barrier filter is used to block out any unwanted or unexpected wavelengths. The fluorescence is usually imaged onto a detector for image digitisation. In wide-field systems, where the whole sample is illuminated, the detector is usually a CCD (charge-coupled device) camera or a more recent sCMOS (scientific compound metal oxide semiconductor) camera. In point-scanning microscopes, the detector is usually a point detector such as a PMT (photomultiplier tube) or APD (avalanche photodiode).

1.2.3 Resolution and Imaging Speed

To understand resolution, we consider the case of an infinitesimally small fluorescent object. When imaged, the fluorescence pattern on the detector is not infinitesimally small. This is because all the information contained in the fluorescence emission cannot be collected by the objective lens which has a finite size and light-collecting power known as the numerical aperture (NA). As a result of this, fluorescent molecules appear as point spread functions (PSFs) rather than individual localisations – which creates blur when several fluorophores are in close proximity (Figure 1.6). The size of the PSF depends on the NA of the objective and the wavelength of light being imaged. In conventional microscopy, which uses objectives with NAs in the range of 0.5–1.5 and wavelengths in the range of 400–800 nm, the width of the point spread function is around 200 nm. Now considering two infinitesimally small objects in close proximity, each will be imaged as a PSF of this width which will begin to overlap, meaning we cannot distinguish between two objects which are less than 200 nm apart.

Biological samples are inherently three dimensional, so in order to capture depth, a sample is split into several thin slices. To avoid physically slicing specimens, optical sectioning systems remove excitation and detection of fluorescence originating from regions outside of the focal plane, essentially removing light which is out of focus from each section. The most common way to do this is to use a confocal microscope. With the aid of optical sectioning, fluorescence microscopy can achieve resolutions of around 500 nm along the optical axis (Figure 1.8). More advanced microscopy systems, such as super-resolution microscopy, can overcome these resolution limits.

Fluorescence microscopy can be used for live-cell imaging over multiple frames, provided a set of conditions are suitably met. A successful live-cell fluorescence microscopy system will minimise phototoxicity to prevent cell damage or death, provide an environment for cells to carry out their metabolic and physiological functions, and sustain a good signal-to-noise

FIGURE 1.8 Optical sectioning splitting a sample into several planes in order to construct a scan. This yields a volumetric representation of the sample.

ratio. The exact parameters of the microscopy system would need to be finetuned for the particular cell type and fluorophore used. However, in general, most cell types will not be able to survive the intensity of light used in the excitation laser for a fixed (dead) sample. Lowering the excitation intensity will counteract this, but reduce the visible distinction between the sample and the background (the contrast). As such, the attainment of temporal resolution comes at the expense of lowering the signal-to-noise ratio.

1.2.4 Types of Fluorescence Microscope

A range of fluorescence microscopy techniques exist for improving the efficiency of particular optical aspects, such as:

- **Wide-field:** In this system, a beam of light illuminates the entire sample to excite all fluorophores. This allows for simple and fast microscopy, but only works well on thin specimens because out-of-focus fluorophores from above and below the imaging plane contribute background. However, these systems are simple and relatively cheap to acquire and operate.

- **Confocal:** A confocal microscope illuminates one diffraction-limited volume of an image at a time and only accepts signal from that region, reducing the level of out-of-focus light. This process inherently takes longer than wide-field, but can achieve higher spatial resolutions especially along the optical axis. Standard confocal microscopy is not appropriate for imaging very fast, dynamic processes. However, there are variations on the method such as the multi-point-scanning Nipkow disk that can address this. Light-sheet microscopes attempt to achieve similar results – the optical sectioning of the sample to allow 3D imaging at high temporal resolution. These are particularly well suited for studying developmental biology (Wan, McDole et al. 2019).

- **Total internal reflection:** When light travels from a medium of high refractive index to a medium of low refractive index at a certain angle, the light will reflect rather than refract. By taking the sample as the medium of low refractive index and a glass coverslip as the medium of high refractive index, total internal reflection microscopy (TIRF) exploits this physical principle (Axelrod, Thompson et al. 1983). Fluorophores within the vicinity of the surface interface can be excited by the resultant evanescent field which only penetrates around 100 nm into the sample. This field decays exponentially in intensity, restricting the excitation of fluorophores to a region much smaller than that achieved by confocal. This yields a greater resolution in the optical axis and a higher signal-to-noise ratio. TIRF is often used to facilitate single-molecule imaging, and although it is restricted to imaging cellular structures close to the coverslip, these frequently include the plasma membrane.

1.3 IMAGING MEMBRANES

1.3.1 Labelling Strategies

The labelling strategy used to stain the plasma membrane depends on the type of dye and system of microscopy used to image the membrane. Generally, small fluorescent dyes which possess a hydrophobic moiety can be applied directly to the sample and will enter the membrane uniformly. However, one of the advantages of fluorescence microscopy is being able to label a particular molecular species. With lipids, this is difficult; however, for membrane proteins, immunostaining with antibodies or the use of fluorescent fusion constructs is possible.

If the cells under imaging are not required to be alive, they can be fixed using a fixative agent – this immobilises cells and restricts the movement of membrane components. This is usually preferable outside of live-cell imaging, as it keeps cell samples stable for longer periods of time and makes staining of samples less prone to the variations seen in live cells. It should be noted though that fixation primarily applies to proteins, and membrane lipids can remain mobile even after fixation (Tanaka, Suzuki et al. 2010).

For live cells, the imaging conditions must be sufficient for cells to sustain themselves. Depending on the cell type, this may necessitate the inclusion of specifically designed culture chambers to regulate the atmosphere – preventing the build-up of carbon dioxide and supplying fresh oxygen. Additionally, any lasers used by the system must have sufficiently low intensity to prevent photodamage or cell death. Fixed (dead) cells do not require such stringent care and can therefore be imaged under harsher experimental conditions. In particular, they do not require regulation of the atmosphere or temperature and can withstand higher laser intensities. Both the sample and the microscope should be free of contaminants and an appropriate laser, dichroic mirror, and barrier filter should be used for excitation and emission detection.

1.3.2 Mapping Membrane Protein Distributions

Membrane proteins are highly specialised, but it is generally agreed that their organisation is equally essential for cell function. For example, neurons require potassium and sodium channels across the entire length of their membrane in order to transmit electrochemical signals, but the exact spatial distribution of these channels (i.e., whether they are heterogeneous, evenly distributed, or spatially random) is as yet unclear. While there is no standardised minimum distance that two molecules can be separated by, the distances between neighbouring proteins on the plasma membrane are typically far below the diffraction-limited resolution of 200 nm. As such, conventional microscopy techniques have, thus far, been unable to probe the spatial organisation of membrane proteins. Current research into mapping the distribution of these molecules requires more advanced techniques, such as super-resolution microscopy (see Chapters 2 and 4).

1.3.3 Measuring Protein and Lipid Diffusion

As discussed, there is mounting evidence of an underlying distribution of liquid-ordered domains in the membrane. These domains, known as lipid rafts, are hypothesised to govern the spatial organisation of proteins in the plasma membrane as a result of each protein's affinity for a particular phase. Generally, the degree of membrane order dictates the diffusivity of molecules, including lipids and proteins, within the membrane. Diffusion is usually slower in the ordered phase as a result of tighter lipid packing and increased rigidity, but tracking the exact diffusive process of individual molecules is no easy feat. Unfortunately, lipid rafts exist at resolutions of sub-200 nm, making them impossible to image via conventional microscopy. Additionally, molecular tracking requires multiple images taken within an acceptably short period of time from one another. In other words, measuring protein and lipid diffusion requires microscopy techniques with high spatial and temporal resolution. This form of microscopy can be difficult to achieve and requires highly specialised equipment, and as such, there is very little data on the exact properties of protein and lipid diffusion (see Chapter 3).

1.3.4 Mapping Membrane Biophysical Properties

Membrane order is but one feature of a cell under the vast umbrella of membrane properties. These properties, such as local viscosity, surface tension, and density, all contribute to healthy cell function. There is a subset of fluorescent molecules known as environmentally sensitive dyes, which alter their emission spectra based on the properties of their environment. In essence, this changes the colour they emit upon excitation. By using multi-channel conventional fluorescent microscopy techniques, such as confocal microscopy, it is possible to attain images at resolutions of up to 200 nm in which the intensity in each channel can be used to quantify spatially averaged membrane properties. In order to probe membrane heterogeneity at resolutions below 200 nm, super-resolution microscopy techniques must be used in conjunction with environmentally sensitive dyes. This also provides data on the spatial distribution of molecules within the membrane, which can help identify correlations between molecular organisation and membrane properties. As it stands, there is still much research to be conducted into interpreting the results of this form of microscopy and determining which dyes can be used to probe particular membrane properties. Additionally, since these dyes require high laser intensities and quickly become photobleached, it is difficult to achieve live-cell imaging over multiple frames without the implementation of specialised dyes (see Chapter 5).

REFERENCES

Axelrod, D., N. L. Thompson and T. P. Burghardt (1983). "Total internal reflection fluorescent microscopy." *Journal of Microscopy* **129**(Pt 1): 19–28.

Clausen, M. V., F. Hilbers and H. Poulsen (2017). "The structure and function of the Na,K-ATPase isoforms in health and disease." *Frontiers in Physiology* **8**: 371.

Gennis, R. B. (1989). *Biomembranes: Molecular Structure and Function*. New York, Springer.

Hancock, J. F., A. I. Magee, J. E. Childs and C. J. Marshall (1989). "All ras proteins are polyisoprenylated but only some are palmitoylated." *Cell* **57**(7): 1167–1177.

Hynes, R. O. (2002). "Integrins: bidirectional, allosteric signaling machines." *Cell* **110**(6): 673–687.

Jacobson, K., O. G. Mouritsen and R. G. W. Anderson (2007). "Lipid rafts: at a crossroad between cell biology and physics." *Nature Cell Biology* **9**(1): 7–14.

Kaiser, H. J., M. A. Surma, F. Mayer, I. Levental, M. Grzybek, R. W. Klemm, S. Da Cruz, C. Meisinger, V. Müller, K. Simons and D. Lingwood (2011). "Molecular convergence of bacterial and eukaryotic surface order." *Journal of Biological Chemistry* **286**(47): 40631–40637.

Lindsley, C. W. (2013). "The top prescription drugs of 2012 globally: biologics dominate, but small molecule CNS drugs hold on to top spots." *ACS Chemical Neuroscience* **4**(6): 905–907.

Lingwood, D. and K. Simons (2010). "Lipid rafts as a membrane-organizing principle." *Science* **327**(5961): 46–50.

Lorent, J. H., K. R. Levental, L. Ganesan, G. Rivera-Longsworth, E. Sezgin, M. Doktorova, E. Lyman and I. Levental (2020). "Plasma membranes are asymmetric in lipid unsaturation, packing and protein shape." *Nature Chemical Biology* **16**(6): 644–652.

Michell, R. H. (1975). "Inositol phospholipids and cell surface receptor function." *Biochimica et Biophysica Acta (BBA) - Reviews on Biomembranes* **415**(1): 81–147.

Mouritsen, O. G. and L. A. Bagatolli (2015). "Lipid domains in model membranes: a brief historical perspective." *Essays in Biochemistry* **57**: 1–19.

Munro, S. (2003). "Lipid rafts: elusive or illusive?" *Cell* **115**(4): 377–388.

Owen, D. M., C. Rentero, A. Magenau, A. Abu-Siniyeh and K. Gaus (2012). "Quantitative imaging of membrane lipid order in cells and organisms." *Nature Protocols* **7**(1): 24–35.

Owen, D. M., D. J. Williamson, A. Magenau and K. Gaus (2012). "Sub-resolution lipid domains exist in the plasma membrane and regulate protein diffusion and distribution." *Nature Communications* **3**(1): 1256.

Rothman, J. E. and J. Lenard (1977). "Membrane asymmetry." *Science* **195**(4280): 743–753.

Schnaar, R. L. (2004). "Glycolipid-mediated cell-cell recognition in inflammation and nerve regeneration." *Archives of Biochemistry and Biophysics* **426**(2): 163–172.

Simons, K. and E. Ikonen (1997). "Functional rafts in cell membranes." *Nature* **387**(6633): 569–572.

Singer, S. J. and G. L. Nicolson (1972). "The fluid mosaic model of the structure of cell membranes." *Science* **175**(4023): 720–731.

Tanaka, K., K. Suzuki, Y. Shirai, S. Shibutani, M. Miyahara, H. Tsuboi, M. Yahara, A. Yoshimura, S. Mayor, T. Fujiwara and A. Kusumi (2010). "Membrane molecules mobile even after chemical fixation." *Nature Methods* **7**: 865–866.

Wan, Y., K. McDole and P. J. Keller (2019). "Light-sheet microscopy and its potential for understanding developmental processes." *Annual Review of Cell and Developmental Biology* **35**(1): 655–681.

Yuste, R. (2005). "Fluorescence microscopy today." *Nature Methods* **2**(12): 902–904.

CHAPTER 2

How to Map Membrane Proteins at the Single-Molecule Level

2.1 SINGLE-MOLECULE MICROSCOPY

In 2014, the Nobel Prize in Chemistry was awarded for the invention and implementation of super-resolution microscopy. Since then, a variety of improvements and alternative methodologies have been introduced, with each technique presenting its own individual advantages and disadvantages. Compared to conventional microscopy, Single-Molecule Localisation Microscopy (SMLM; Lelek, Gyparaki et al. 2021) can achieve a tenfold increase in resolution – this dramatic increase is owed to the use of fluorescent molecules, embedded within the membrane, which are subjected to wide-field excitation (in which the whole sample is illuminated by light of a specific wavelength) and blink sequentially over a range of frames. Molecules are then computationally localised from the resulting series of images to reconstruct the super-resolution image, so that qualitative and quantitative information can be extracted from which biological analogies are drawn. Unlike the 2D or 3D images derived from conventional microscopy, SMLM outputs point clouds – these are typically 2D or 3D, although often presented with additional information regarding localisation precision. Point clouds are characterised as big data and may contain thousands, if not millions, of individual localisations. As such, it is highly inefficient to attempt to interpret the resulting data manually, and quantification

DOI: 10.1201/9781003273745-2

19

methods, such as cluster analysis, must be used. In the following sections, a selection of the most popular SMLM techniques will be explored, including Photoactivated Localisation Microscopy (PALM; Betzig, Patterson et al. 2006), Point Accumulation for Imaging in Nanoscale Topography (PAINT; Schnitzbauer, Strauss et al. 2017), Stochastic Optical Reconstruction Microscopy (STORM; Rust, Bates et al. 2006), and Direct STORM (dSTORM; Heilemann, van de Linde et al. 2008). Each variety presents its own advantages, disadvantages, and range of parameters, depending on the type of analysis desired.

2.1.1 Photoactivated Localisation Microscopy (PALM)

PALM exploits a specific variety of fluorescent proteins known as photoactivatable fluorophores. These fluorophores can be switched to an active state via UV illumination before becoming irreversibly photobleached (total loss of fluorescent properties). Photoactivation is typically achieved using UV illumination and using very low UV powers, where only a small subset of fluorescent molecules will be activated. These can be imaged and will appear as a sparse set of PSFs on the camera. These can be recorded and the PSFs can then be fitted to find their centre, which represents the position of the fluorescent molecule. The fluorophores are then bleached and a new subset of fluorophores activated and imaged. Through this iterative process, all fluorescent molecules in the sample can eventually be activated, imaged, and localised. Examples of fluorescent proteins commonly used for this application include PS-CFP2 and mEos3.

A variant of PALM known as spt-PALM is used in single-molecule tracking, which records information about the dynamics of molecules within a region of interest (ROI; Manley, Gillette et al. 2008). This information can come in one of two forms, depending on whether the information on dynamics is stored under the Lagrangian or Eulerian method. In the case of the Lagrangian method, each individual molecule's dynamics are tracked across the ROI, e.g., by estimating the diffusion coefficient from mean-squared displacements or probing transport states for each molecule. On the other hand, the Eulerian method characterises dynamics across the entire region by inferring local properties that might impact molecule dynamics from the behaviour of molecules passing through the region. This can be used to generate maps of diffusivity or energy potentials. In either case, it is imperative to be mindful of localisation errors, which bias the apparent molecular dynamics observed.

2.1.2 Point Accumulation for Imaging in Nanoscale Topography (PAINT)

Unlike many of its counterparts, PAINT does not actually require photo-switching, but instead exploits fluorophores which diffuse freely and then become immobilised upon binding to a target. The most popular form of PAINT, known as DNA-PAINT, achieves this immobilisation via hybrid-isation of DNA strands. These strands, known as "imager strands," are dye-labelled oligonucleotides that bind specifically to their complements, known as "docking strands," which are attached to their target. Dyes or dye-labelled ligands diffuse freely until they reach their target molecules and bind permanently or transiently. These dyes can diffuse rapidly across a range of pixels during the time taken to acquire one frame – as such, they will appear as blurred streaks and not be localised. However, once they are bound, the dyes will be fixed, and a PSF can be fitted. PAINT is not impacted by photobleaching as stable, bright fluorophores can be used and replenished in the sample.

2.1.3 (Direct) Stochastic Optical Reconstruction Microscopy (STORM)

In STORM, target molecules are labelled through immunofluorescent labelling – that is, by labelling proteins with antibodies that have been con-jugated to fluorescent dye molecules (Figure 2.1). To achieve the required blinking behaviour, dye pairs such as Cy3 and Cy5 are often used, with Cy5 acting as the primary fluorophore, which can be reversibly photoswitched, while Cy3 used to facilitate the switching of Cy5. This switching cycle can occur thousands of times before photobleaching ensues. However, the requirement to use two dyes in close proximity makes the sample prepara-tion challenging. dSTORM is an extension of STORM which makes use of a single photoswitchable fluorophore that can be reversibly switched between an active and an inactive state by irradiation with light of differ-ent wavelengths. This means that dSTORM can image cellular structures with a resolution of around 20 nm, without the need for a facilitating mol-ecule. This is achieved by using chemical buffers that prolong the life of naturally occurring fluorescent dark states. dSTORM also makes use of immunofluorescent labelling, but in this case, fluorophores are induced into an inactive dark state by laser illumination and then stochastically return to their fluorescent ground state. The most popular fluorophore for dSTORM is Alexa 647, which displays high photon yield and consistent blinking.

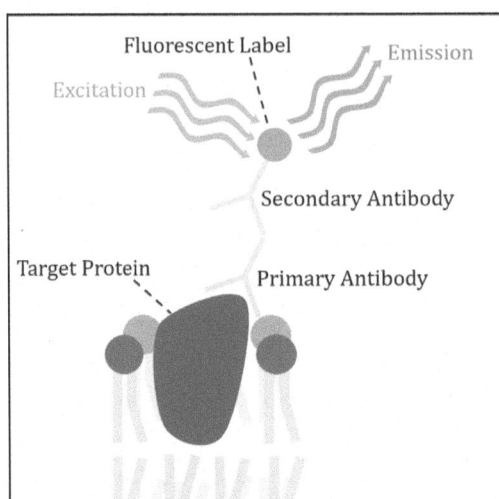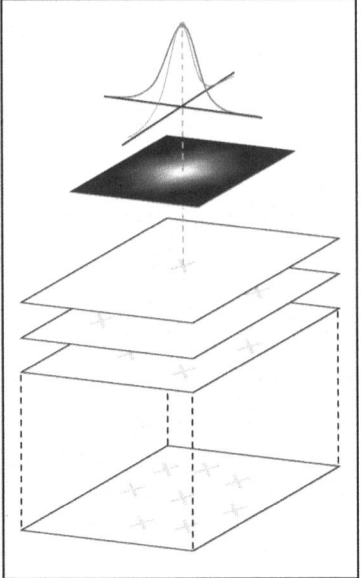

FIGURE 2.1 Workflow of STORM SMLM. Fluorophores are attached to membrane proteins by immunofluorescence labelling. Individual PSFs are recorded over several frames, with Gaussian fitting applied to each, and combined together to give a point cloud.

2.2 IMAGING

The range of experimental and analytical parameters required for versatile SMLM imaging far exceeds that of conventional microscopy. To achieve the desired imaging outcome, parameters and methodologies must be optimised. With the addition of stochastic blinking and the increase in noise it brings, this makes even the simplest of SMLM experiments difficult to reproduce. In the following sections, we discuss the considerations that must be made prior to conducting SMLM and encourage careful documentation of any methods and protocols carried out during the experiment – from labelling and imaging to analysis and interpretation. Any data derived and software used should be open source and publicly available for ease of reproducibility. The data derived from SMLM is inherently large, so at the very least information on the localisation data should be released (especially if no other data can be made available) – this should include the spatial coordinates, frame number, and photon count of each localisation. The protocols that must be stated are: the fluorescent labelling technique (including type of fluorophore, emission wavelengths, antibody concentration, and method of protein tagging), the method of fixation, the type of microscopy used, and any microscope parameters (laser power, excitation wavelength, type of camera, pixel size, exposure time, magnification, numerical aperture and immersion medium used) as well as any filtering or post-processing software (and any parameters they incorporate).

2.2.1 Labelling Methods

The labelling method used in SMLM will impact both the labelling density and the linkage error (the distance between the fluorophore and the target molecule) and therefore determines whether the target molecule is precisely localised. Furthermore, only certain types of labelling methods are compatible with live-cell imaging. As such, the choice of labelling method will impact the maximum achievable resolution of the SMLM experiment. There is a range of methods to bind fluorophores to target molecules in SMLM. Typically, one of the following methods will be used:

- **Genetic fusion:** A fluorophore may be fused directly to a target protein or the target protein may be fused to a tag that binds to synthetic dyes and can then be introduced to the sample. Genetically encoded fluorescent proteins are among the most popular for live-cell imaging. Genetic fusion is typically achieved by transfection – a process in which an expression construct (such as a plasmid or virus designed for gene expression in cells), containing genetic material which encodes the target protein genetically fused to a fluorescent protein, is introduced into the cell (typically via electroporation). However, in transient transfection, the number of target proteins expressing the fluorescent protein can vary, and in some cases, fluorescent proteins may interrupt the normal function of the target protein, disrupting normal cell function. In the ideal scenario, each target protein will carry precisely one fluorescent protein that does not impact its function. Similarly, protein tags can be genetically fused to target proteins, which covalently bind to their ligand, and this ligand can be directly coupled to a fluorophore. Since these tags are small molecules, there is very little linkage error. Additionally, they can be used for live-cell imaging.

- **Immunolabelling:** This technique exploits the tendency for antibodies to bind to antigens. In this particular case, the antibody would be tagged with a fluorescent dye, creating what is known as a dye-conjugated antibody. This method is typically used in conjunction with synthetic dyes, as they cannot be genetically encoded onto target proteins, and therefore must be coupled with a compound that can bind to the target protein. In immunolabelling, that compound is an antibody that binds specifically to the target protein, an antigen. In direct immunolabelling, one primary antibody carries the synthetic dye and binds to the target molecule. In indirect immunolabelling, a secondary antibody carrying a synthetic dye binds to an unlabelled primary antibody, which then binds to the target molecule. The advantage of indirect immunolabelling is that multiple secondary antibodies carrying synthetic dyes can bind to the primary antibody, therefore amplifying the fluorescent signal. Additionally, primary antibodies labelled with synthetic dyes have a lower epitope binding efficiency. Therefore, indirect immunolabelling can reduce the proportion of free synthetic dyes, and therefore reduce background noise. However, since antibody complexes are particularly large molecules (measuring around ~10 nm per antibody), immunolabelling carries a high linkage error, often leading to a reduction in localisation precision. Immunolabelling can only be carried out on live cells if the target proteins are extracellular (which membrane proteins are) and do not impact the target molecule's function.

- **Direct binding:** Synthetic dyes can undergo direct binding to specific structures on the plasma membrane if they are coupled to certain small dye-conjugated proteins or drugs. As these are inherently small molecules, the linkage error arising from such a process is very small. Unfortunately, these labels often impact biophysical function, making them unsuitable for live-cell imaging.

2.2.2 Fluorescent Dyes and Photophysics

Fluorophores are selected for their high absorption cross section and high quantum efficiency (the number of photons the fluorophore emits almost matches the number of photons it is excited with), which makes them particularly bright and therefore easy to distinguish from background signals and noise.

There are four categories of fluorophores used in SMLM:

- **Photoswitchable fluorophores:** These fluorophores can undergo multiple, reversible transitions between active and inactive states.

- **Photoactivatable fluorophores:** These fluorophores can be switched from an inactive to an active state irreversibly. This process is typically stochastic but can be activated by UV light.

- **Photoconvertible fluorophores:** These fluorophores can switch from one colour (known as a spectral state) to another irreversibly. This process is brought about by UV irradiation. These fluorophores can then be imaged at different wavelengths. This is useful, as one spectral state (typically the one with shorter wavelength) can be used to focus the sample. Then, the fluorophores can be photoconverted to the second spectral state and activated.

- **Spontaneously blinking dyes:** Unlike the previous examples, these dyes are not controlled by irradiation. Instead, they employ a reversible pH-dependent reaction in order to activate. This means that fluorescence is controlled by the pH of the buffer, which can be difficult to alter. However, these dyes have a high photon yield and are suitable for live-cell imaging.

These categories include both synthetic dyes, which have a higher photon count and allow for greater localisation precision and shorter imaging times, and fluorescent proteins, which are better suited to live-cell imaging than synthetic dyes, but are more prone to premature bleaching. When selecting an appropriate fluorophore, it is important to consider the following factors:

- **Photon output:** Is the number of photons output by the fluorophore during emission. The higher the photon output, the more accurate the localisation.

- **Duty cycle:** Is the period of time in which the fluorophore remains in its active and inactive states. This should be as close to the exposure time of the camera as possible.

- **Switching cycle:** Is the number of times the fluorophore can switch between active and inactive states. Reducing this will decrease the chance of multiple blinking occurring, but it may also mean that some fluorophores are not localised.

Living cells can be highly responsive to light and some might be destroyed by phototoxicity if the intensity of light is too great. As such, dead cells can be imaged at a far higher resolution than live cells. That said, it is still possible to achieve resolutions as low as 50 nm in live samples. In live-cell imaging, fluorophores must photoswitch quickly to ensure rapid turnover and sampling before cell death, and cellular physiology must not be affected by the imaging conditions, buffer solution, or phototoxicity.

2.2.3 Excitation, Emission, and Blinking

The diffraction limit of an optical system was first quantified by Ernst Karl Abbe in 1873 and is described mathematically as follows:

$$d = \frac{\lambda}{2NA},$$

where d is the resolution limit, λ is the wavelength of light, and NA is the numerical aperture of the objective lens. As such, the diffraction limit is proportional to the wavelength of light and inversely proportional to the numerical aperture. Visible light is constrained to a minimum of 400 nm, meaning that even with an idealised aperture of around 1.5, the diffraction-limited resolution cannot breach the barrier of 200 nm.

Collectively, overlapping PSFs can be difficult to identify and make localisation almost impossible. In most biological samples, target molecules will not be spatially distinguishable due to the high density of membrane proteins. However, by sequentially exciting small subsets of fluorophores over several frames, each individual image can be probed and the exact spatial coordinates of the fluorescing molecules can be identified with greater precision. To achieve this, SMLM exploits the stochastic photoswitchable property of fluorophores. This process allows fluorescing molecules to switch to an active "bright" state and then back to an inactive "dark" state. Photoswitching is a probabilistic process, and under normal conditions, it would be unpredictable; however, it is possible to adjust switching probabilities using techniques such as laser irradiation (among others). This switching of fluorophores results in a phenomenon known as blinking, in which fluorophores illuminate sporadically across thousands of image frames. Under optimal conditions, it is possible to separate the fluorescent emissions of small subsets of distinct molecules in time and therefore avoid overlapping PSFs entirely. However, this requires that only a small number of molecules fluoresce in each frame and that every molecule fluoresces in at least one frame without multiple blinking, which is not often the case. Once all frames have been acquired, each image is processed individually to detect and localise all active molecules.

Ultimately, each localisation is aggregated into a final image known as the super-resolution image (Figure 2.1).

The more excitation and emission cycles a fluorophore experiences, the greater the chance of its own light-induced destruction, causing the fluorophore to lose its ability to fluoresce. This process is known as photobleaching and it is believed to occur because excited molecules are more likely to react with other molecules in their environment, particularly oxygen, causing them to lose the ability to emit. Typically, organic dyes can emit between 10^5 and 10^6 collectable photons before becoming photobleached, while fluorescent proteins can emit between 10^4 and 10^5 collectable photons. While certain anti-photobleaching solutions do exist, they act by reducing the amount of oxygen available to the fluorophores and can often be toxic to living cells. Each fluorophore presents its own advantages and disadvantages which will affect the spatial and temporal resolution of the system. The fluorophore and optical system are selected depending on the desired result of the analysis, for instance, 2D or 3D imaging, one or multiple colours, fixed or live cells, labelling strategies, sample preparation (e.g., whether and how samples are fixed, the buffer that is used to promote photoswitching), method for acquiring imaging sequences, laser power, and localisation software used (Dempsey, Vaughan et al. 2011).

2.2.4 Imaging Hardware

There are two main components to the hardware required for SMLM, namely: a wide-field microscope, equipped with standard continuous wave lasers for excitation and activation, and a camera capable of detecting single molecules. The most commonly found cameras for such an experiment are:

- Electron-multiplying charge-coupled device (EM-CCD): This device can detect single photons with negligible read-out noise, making them particularly applicable for low photon counts.

- Scientific complementary metal-oxide semiconductor (sCMOS) camera: This camera has a higher frame rate and larger field of view that EM-CCDs. This will typically have a similar signal-to-noise ratio (SNR) for bright dyes to EM-CCDs. However, this is less sensitive to weak signals and produces read-out noise that requires correction.

There are several considerations that must be made when selecting a detector. The pixel size derived from the detector should be approximately equal to the standard deviation of the PSF (Thompson, Larson et al. 2002). Any camera used in SMLM can be impacted by several sources of noise such as shot (Poisson) noise, which arises as a result of low photon count and is therefore inevitable for most fluorophores, and dark current noise, which arises from electrons created during the imaging process (often as a result of thermionic emissions) that are independent of the signal captured by the detector but still counted as signal. Dark current noise can be reduced by incorporating a cooling system into the camera. Cameras can also be subject to read-out noise (the noise arising from the amplifier which converts charge from electrons into voltage), although this is often negligible, and multiplicative noise (the noise arising from stochastic amplification of photoelectrons), which can be reduced by incorporating a low amplification gain.

2.2.5 3D Methods

There are several approaches to achieving SMLM in three dimensions. One such method is to introduce a cylindrical lens into the SMLM setup which produces an astigmatic PSF (Figure 2.2). Additional optical devices can be

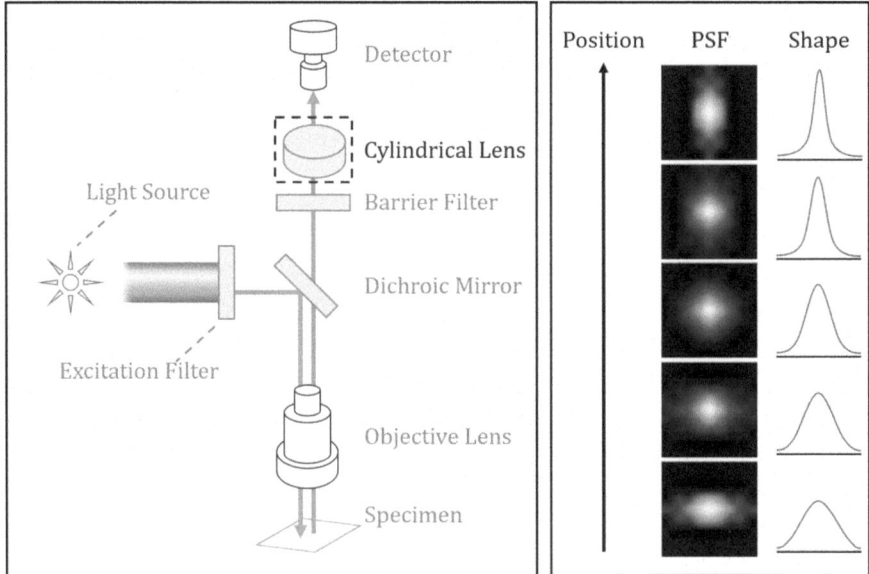

FIGURE 2.2 Updated optical diagram of a fluorescence microscope with the addition of a cylindrical lens, which produces astigmatism. The effect of astigmatism on the shape of a PSF is shown at different axial variations.

introduced to further alter the shape of the PSF – this process is known as PSF engineering (Huang, Wang et al. 2008). In general, as the axial coordinate of the molecule of interest changes, so will the PSF, meaning that the axial coordinate can be estimated from the shape of the PSF. For example, 3D STORM has been achieved by incorporating a cylindrical lens to introduce astigmatism to the image. As a result, the fluorophore appears to be ellipsoidal, and by fitting the image with 2D elliptical Gaussians, the peak widths of the x, y coordinates could be determined and the z-coordinate could be calculated. Other techniques, such as those used in 3D PALM, implement biplane detection – this involves using a beam splitter to split light into a shorter and longer path, forming two detection planes for z-position determination (Juette, Gould et al. 2008).

2.3 IMAGE PROCESSING

Super-resolution images must be processed computationally, so the quality of the resulting images will depend on the image processing methods used. Image processing can be decomposed into three stages: single-molecule detection, single-molecule localisation, and super-resolution image rendering. These stages will be explored over the following sections, alongside some immediate considerations that should be made for correcting image artefacts.

2.3.1 Filtering

During single-molecule detection, each frame is analysed to detect the active emitters. Pre-processing techniques such as rolling ball algorithms, wavelet filtering, or difference of Gaussian algorithms can be used to remove the heterogeneous background and make the active emitters appear more pronounced in the image. This results in a background-corrected image, in which detection can be undertaken by a number of methods (Figure 2.3). One such method, although somewhat primitive, is to simply identify all local maxima over the intensity profile of the image. A more advanced and accurate method is to actively locate PSF-like patterns in the image by Gaussian fitting (that is, fitting Gaussian functions to the point cloud) or by determining the correlation between the image and a model PSF (this will depend on the expected shape of the PSFs, which may differ significantly in 3D SMLM, but typically a Gaussian approximation of the Airy function will suffice). The resulting correlation image can then be binarised by thresholding. In any case, the result of this stage will be a binarised image – the idea is that any pixel regions which show up in the

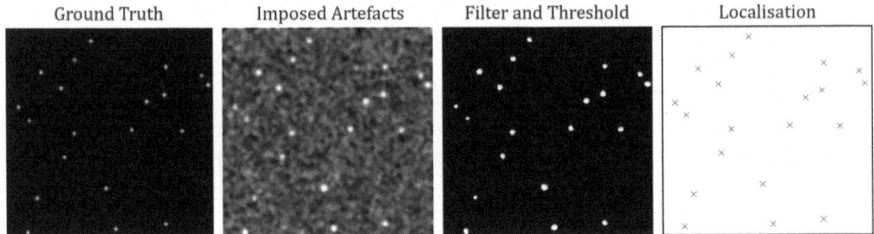

FIGURE 2.3 The results of applying a filtering algorithm, used to negate the effects of imaging artefacts.

binarised image are likely to contain molecules. Each detection algorithm has its own drawbacks and benefits, but almost all will lose accuracy if the images are especially noisy – this loss of accuracy typically comes in the form of an increased number of false positive and false negative detections. In images with a particularly low SNR, these errors are unavoidable. In general, the higher the SNR, the better detection will be.

2.3.2 Drift Correction

When imaging at the nanoscale, variations in sample position can produce seemingly large localisation offsets or blurring in the reconstructed image. This phenomenon is known as drift, and it is one of the most common artefacts in SMLM. Drift can typically be in the range of tens of nanometres and is often longer for slow imaging systems. Fixed luminescent beads can be added to samples to calibrate drift during post-acquisition analysis.

2.3.3 Localisation

In this stage, the pixel regions identified in the filtering step are analysed to determine the subpixel coordinates of each molecule. Before delving into the methodologies behind localisation, it is imperative to understand two notions which may impact localisation: these are precision and accuracy. In general, the variance defines the precision of an algorithm, while the bias defines the accuracy. An algorithm has high precision if it consistently provides similar coordinates for different images of the same molecule; however, this may not necessarily be an accurate representation of the coordinates of the molecule. On the other hand, an algorithm has high accuracy if it tends to determine the correct position of the molecule on average; however, the variance between these localisations may be quite large, meaning points are dispersed and therefore coordinates can differ. The ideal algorithm has both high precision and high

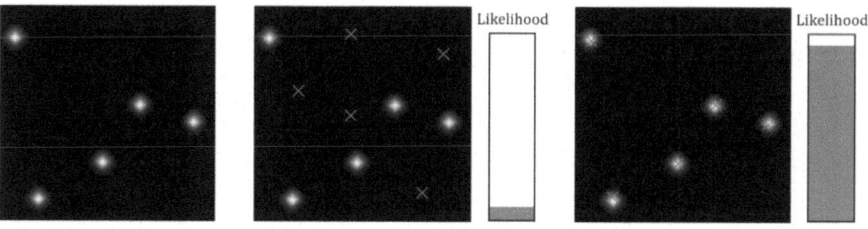

FIGURE 2.4 The process of the MLE algorithm. The closer the localisations to the centres of the PSFs, the greater the likelihood.

accuracy – consistently localising in the same small region which accurately represents the true location of the molecule.

The most popular algorithm for localisation is Maximum Likelihood Estimation (MLE; Figure 2.4), which computes the coordinates of each molecule for which the likelihood of obtaining the observed image is at its highest. This process is usually undertaken incrementally, with coordinates initially placed arbitrarily in pixel regions across the image and then offset slightly at each iteration until the likelihood cannot be increased any further. Each MLE algorithm makes assumptions about the PSF shapes, background, and noise, which ultimately will impact its performance. As the SNR increases, MLE approaches the lower bound of precision (known as the Cramér–Rao lower bound or CRLB) both theoretically and experimentally, although achieving the limit in practice requires an accurate model of the PSF. Typically, a Gaussian or Airy pattern will be used (with the latter providing greater accuracy), but more advanced functions from the field of optical theory can be incorporated. It is possible to calibrate PSF models before each experiment by cubic spline fitting or phase retrieval algorithms on images of sub-diffraction fluorescent beads. It has been shown that experimentally derived PSFs outperform idealised model PSFs. As it is an iterative process, computation times for MLE often exceed that of faster, although less precise, algorithms such as intensity-weighted centres of mass; however, as graphics processing units have become more advanced, MLE algorithms can now analyse SMLM images in real time. Once each detected molecule has been assigned a localisation, post-processing techniques can be used to filter out suboptimal localisations, such as those in which the model PSF does not fit the image or the CRLB falls above some pre-determined threshold. This form of filtering can improve the average localisation precision but may not necessarily improve resolution as removing localisations can jeopardise sampling.

Once detection and localisation have been undertaken on all diffraction-limited images, all images are concatenated into one and rendered as a super-resolution image. In this case, bin (pixel) sizes are usually taken to be equal to the precision with the intensity of that pixel corresponding to the number of localisations identified in its bin. There are other methods of rendering the super-resolution image such as applying a weighting to localisations from higher-intensity PSFs or employing density estimation. Determining the exact resolution of a super-resolution image can be troublesome, as there are a number of different techniques that can be employed, and it is not immediately obvious which to use. One simple method is to take the spatial standard deviations of localisations in small clusters, each assumed to pertain to a single molecule; however, this ignores potential localisation biases and linkage errors. One of the most effective methods for estimating resolution of an imaging technique is to apply that technique to a data set where the molecules imaged are at a known distance from each other, e.g., in nuclear pores. If multiple images of the same complex have been taken, then further information about the molecular structure of the complex can be derived by superimposing and averaging SMLM images after translations and rotations. This approach is known as single-particle reconstruction, and although the resolution of a reconstructed image still depends on localisation precision, this form of averaging allows us to probe structural features with a precision limited by the number of images of the complex rather than the single-molecule localisation precision.

There is an expansive set of software packages for analysing SMLM data available online, although most require either an existing image analysis software, such as ImageJ or Fiji, or a programming language capable of incorporating image analysis, such as Python, R, or MATLAB (Ovesný, Křížek et al. 2014).

2.3.4 Blink Correction

Blink correction is the name given to a range of post-processing techniques which aim to counteract the effect of multiple blinking. Since multiple blinking will often be expressed across a series of consecutive frames, one such technique is to merge these nearby localisations across the frames in which they appear and conglomerate them into one single, spatially averaged localisation. This yields a more precise localisation without compromising sampling density. However, there are more specialised software available for dSTORM and PALM imaging (Bohrer, Yang et al. 2021; Jensen, Hoh et al. 2022).

2.4 IMAGE QUALITY METRICS

Ascertaining quantitative data about the quality of an image can be just as important as interpreting the image itself. In the following sections, we will explore the existing measurements of image quality and denote methods for determining them.

2.4.1 Density and Localisation Precision

In the absolute absence of bias, the precision of an imaging system is limited by the CRLB, which is given by

$$\sigma_{loc} \geq \frac{\sigma_0}{\sqrt{N}},$$

where σ_{loc} is the precision (standard deviation of errors in estimated coordinates), σ_0 is the standard deviation of the PSF, and N is the number of photons collected by the camera. It is clear in this instance that there is a direct proportionality between the precision and the SNR. However, this is a best-case scenario, which does not consider other factors that impact precision, such as the read-out noise, multiplicative noise, background signal, dipole orientation, the non-Gaussian shape of the PSF, and finite pixel size, all of which increase the precision limit. That said, amendments to the CRLB which take these factors into account do exist, e.g.,

$$\sigma_{loc} \geq \sqrt{\left(\frac{12\sigma_0^2 + a^2}{12N}\right)\left(\frac{16}{9} + \frac{8\pi\sigma_0^2 b^2}{a^2 N^2}\right)},$$

where a is the pixel size and b the background intensity. The concept of the CRLB is useful in SMLM as it offers a fundamental limit to which all localisation algorithms can be compared (Thompson, Larson et al. 2002).

2.4.2 Fourier Methods and Resolution

There are a number of methods for determining the exact resolution of an image derived from SMLM, the most prevalent of which involves Fourier analysis. This involves constructing a function known as the Discretised Fourier Transform (DFT) for an image. Once calculated, the DFT can be represented as an image, known as the Fourier space, which is typically visualised as an approximately circular white light in the centre of a dark space. The exact mathematical notation of the DFT for an $n \times n$ pixel image is given by

$$F(x,y)=\sum_{k=1}^{n}\sum_{l=1}^{n}f(k,l)\,e^{-2\pi i\left(\frac{kx+ly}{n}\right)},$$

where $F(x, y)$ is the intensity value of the Fourier space in pixel (x, y), $f(k, l)$ is the intensity value of the original image in pixel (k, l), and $i = \sqrt{-1}$ is the so-called imaginary number. It should be noted that the Fourier transform can be easily computed using Fast Fourier Transform (FFT) software packages, which exist for most programming languages and image analysis software. Taking this image at the log scale can highlight the central peak and the surrounding features. Then, the radial average can be determined by taking the integral of all pixels at a given radius from the centre. Once this has been calculated against each possible radial value, the radial averages can be plotted against the inverse of the length they were determined at. The location on the graph at which information content disappears into background noise serves as an estimate for the resolution that the system can achieve. This process is known as Fourier Spectrum Analysis, and its main advantages are that it does not depend on the SNR and can determine a resolution for the image as a whole rather than just isolated regions.

2.4.3 Artefact Detection

An imaging artefact refers to a feature of the produced image which was not present in the original imaged object. At such a sensitive scale, these are common in SMLM, and they can therefore bias quantification and interpretation of results (Culley, Albrecht et al. 2018; Marsh, Costello et al. 2021). Measures should always be taken to reduce the impact of artefacts in the data before conducting analysis. While the likelihood of artefacts can be minimised by following the SMLM protocols exactly, they are often inevitable and require computational post-processing to mitigate. The most common types of artefacts are discussed in the following section.

2.5 POTENTIAL IMAGE ARTEFACTS

In an ideal SMLM experiment, each molecule will be fluorescently labelled by precisely one fluorophore, which will be localised in precisely one of the multiple frames and emit a large number of photons relative to the background. In reality, these conditions are difficult, if not impossible, to achieve. This results in phenomena which are not actually present in the data but appear as a result of the procedure, known as artefacts. In this section, we will discuss the potential image artefacts that may arise while undertaking SMLM and disclose methods of overcoming them.

2.5.1 Multiple Blinking

Many frames are recorded throughout the process of SMLM, and the same molecule can be localised across multiple frames (Figure 2.5). This results in one of the most severe imaging artefacts, known as multiple blinking. This phenomenon can result in a significant increase in density across areas of the ROI and may even produce pseudoclusters in which a single fluorophore is localised so many times in close proximity that it is detected as an entire cluster. Multiple blinking is a stochastic process, meaning that the experimental design can only be optimised so much to prevent it. However, there are computational methods, such as distance distribution correction and cluster filtering, which were developed to combat blinking in post-processing.

2.5.2 Localisation Precision

The degree of uncertainty surrounding the position of a molecule is known as its precision. In general, precision is a result of the uncertainty in determining the centre of the PSF (Figure 2.6). The precision is therefore proportional to the width of the PSF and inversely proportional to the square root number of detected photons. Such localisation precision is a measure of the spread of possible spatial positions at which the fluorescing molecule could be localised; in other words, it is an estimate of the distance between the localisation of the molecule and its actual position. Localisation precision is limited by the SNR and not by the wavelength of light or pixel size,

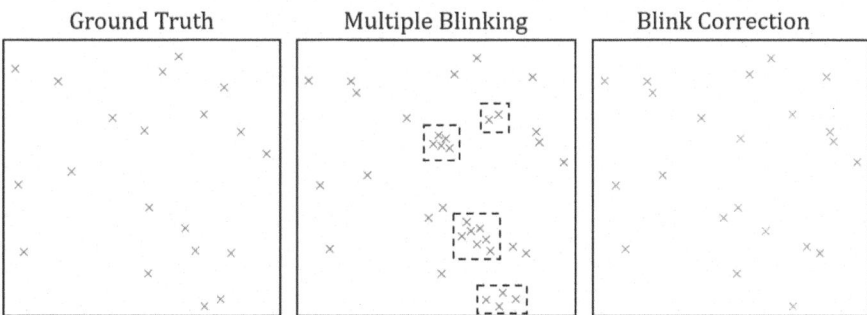

FIGURE 2.5   The impact of multiple blinking and the results of applying blink correction.

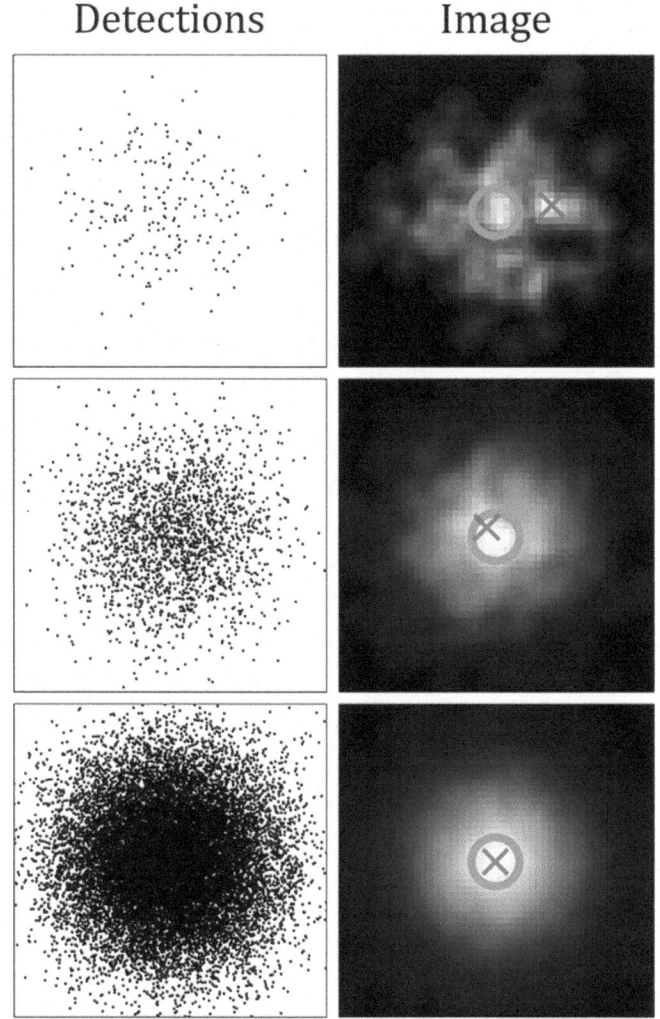

FIGURE 2.6 The impact of photon count on SNR and the perceived PSF. The higher the photon count, the more easily identifiable the PSF and the greater the precision of the localisation.

and as the photon count increases, so does the SNR. Therefore, as the photon count increases, so does localisation precision.

Variations in precision may also arise from computational analysis. For instance, localisation bias can occur in contexts when the MLE algorithms do not converge correctly. This is particularly apparent in cases when the model PSF is a poor fit for the actual PSF. This often results in super-resolution images which appear to have been intersected by a grid-like pattern.

The most effective method of improving localisation precision is to amend the microscopy technique used. In samples, cell membranes can be found close to the coverslip. This means that total internal reflection fluorescence (TIRF) or highly inclined and laminated optical sheet (HILO) illumination can be used to reduce background, increase SNR, and improve localisation precision. TIRF is a microscopy technique which incorporates a strongly inclined laser beam that is reflected by the coverslip sample interface. This leaves only a thin (~200 nm) layer from the sample illuminated and thus reduces the background. HILO is a microscopy technique in which the laser beam enters the sample at a sharp angle, thus reducing the background when imaging at a distance from the coverslip. More generally, low-intensity excitation radiation and long exposure times can minimise photobleaching and improve precision in most experiments.

2.5.3 Drift

When imaging samples at the nanoscale, even slight, unavoidable variations in sample position can translate to relatively large localisation offsets or blurring of the reconstructed image in results (Figure 2.7). This phenomenon is known as drift, and it is one of the most common artefacts in SMLM. Thankfully, it is also fairly easy to rectify, with both experimental and computational procedures suited to handle it. One such experimental method is to incorporate fiducial markers into the sample. These typically come in the form of fluorescent beads which, theoretically, remain fluorescent for the entire imaging process and are therefore visible in all frames acquired. The dynamics of these fiducial markers across each pair of frames serves as a good approximation for the drift of the entire sample, so the true position can be estimated by subtracting this drift. Some imaging systems automatically update the position of the sample using piezoelectric actuators in real time. Although this is more computationally demanding and difficult to implement than post-process drift correction, it has been shown to provide a greater reduction in the error of drift.

If the incorporation of fiducials in not practical, then computational methods may suffice instead. The most common computational method of drift correction is cross-correlation, which approximately measures the change in position of identifiable structures present in multiple frames. However, this method requires that identifiable structures are present in the data, which is not always the case and is unlikely to be as accurate as the use of fiducials. Drift can also occur in the axial direction, so it is important to incorporate 3D drift correction methods even in 2D SMLM.

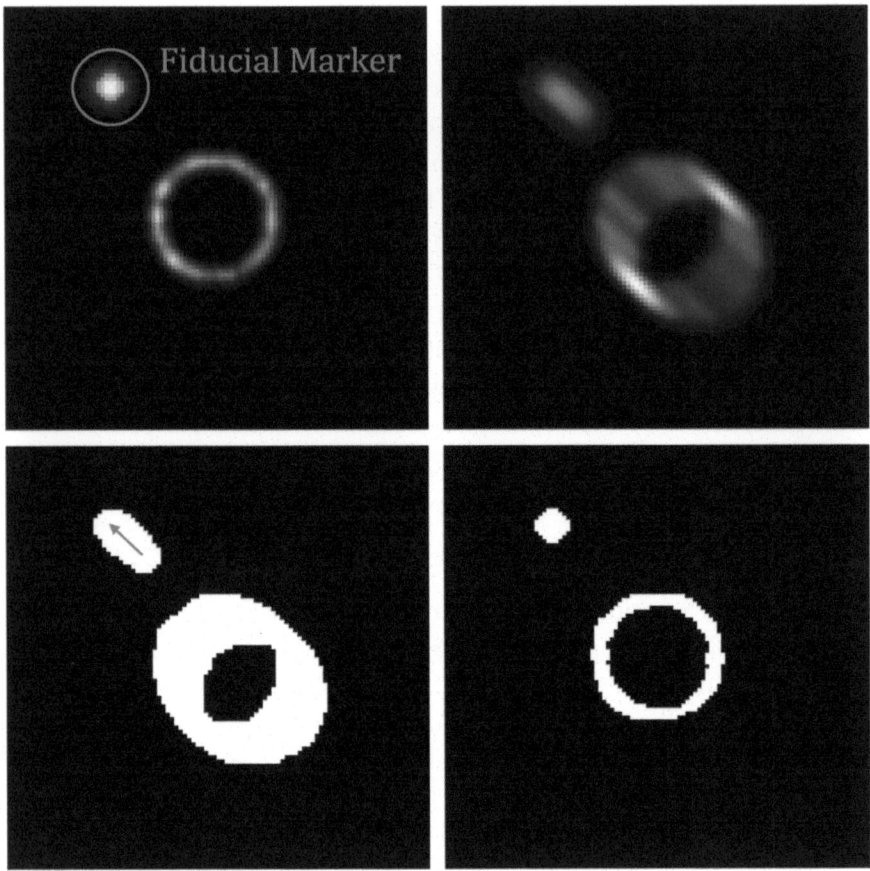

FIGURE 2.7 The impact of drift and correction with fiducial markers. As a result of drift, a ring like structure appears stretched in the acquired image. By tracing the movement of a fiducial marker, localisations can be offset to their original position.

2.5.4 Over-Density

In the ideal acquisition, the number of active fluorophores in each frame will be maximised while the overlap of PSFs is minimised, thus minimising the acquisition time. Unfortunately, it is difficult to avoid overlapping PSFs entirely, especially in dense domains in the ROI or in cases when a high activation probability is required for fast imaging. This can cause problems during the localisation stage, as most localisation algorithms treat overlapping PSFs as the PSF of a single molecule and compute the single corresponding localisation as being somewhere in between the

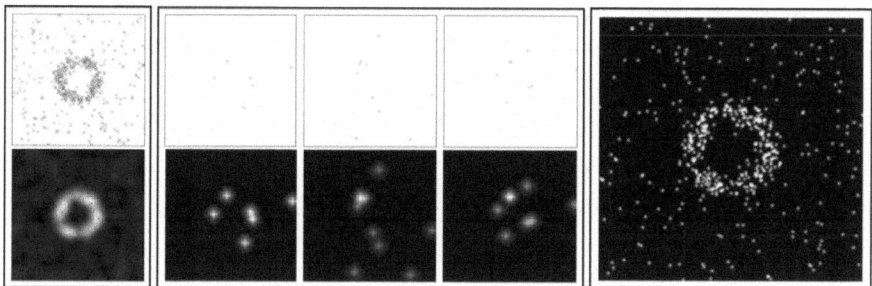

FIGURE 2.8 The impact of imaging several overlapping PSFs, compared to PSFs taken stochastically in separate frames.

actual location of the molecules. Ultimately, this creates artificial blurring in the reconstructed image, particularly at the intersections of imaged structures (Figure 2.8).

The density of active fluorophores is often subject to the type of SMLM being performed and the type of fluorophore used. For instance, the density of photoswitching dyes is controlled by the trade-off between excitation and activation laser intensities. On the other hand, the density of temporarily binding dyes is dependent on the dye's binding affinity and concentration. These parameters can be determined manually; however, it is possible in some cases to automate this process using control software that implements a feedback loop to tune the active fluorophore density.

One of the most common ways of dealing with overlapping PSFs is to filter out localisations in which the observed PSF does not fully match the model PSF. However, this requires a suitable degree of error as rigorous filtering can delete most localisations, reducing sampling and resolution while offsetting density in the reconstructed image. More advanced methods incorporate multi-emitter fitting algorithms, which are specifically designed to interpret PSF overlaps or calculate differences between consecutive frames which are then processed to localise single molecules.

2.6 QUANTIFICATION

SMLM is unique in the sense that it offers information about the number of target molecules, their density, and their distribution across a region of interest. This versatility means we can derive information about the localisations of millions of proteins (if not more) embedded within a plasma membrane at the nanoscale. However, since the point clouds derived from such analysis contain so many points, it can be difficult to ascertain information from the data by eye. As such, computational methods have been developed to quantify and summarise the spatial distribution of points with minimal human intervention. Throughout the following chapter, we will explore methods of extracting the important data from SMLM, choosing and applying quantification methods, such as cluster analysis, and interpreting the results.

2.6.1 The Nature of SMLM Data

The typical output of SMLM comes in the form of a table of coordinates. At the very least, this table will include the coordinates of each localisation, but might also feature the photon count, localisation precision, and frame number, among other variables. By extracting the coordinates of each localisation, this data can be most easily visualised as a scatter plot or point cloud.

It is essential to ensure that all the necessary post-processing has taken place prior to cluster analysis and quantification. At the very least, localisation artefacts should be appropriately handled so that they do not impact the density of the resulting point cloud and produce false positive or false negative clustering. Arguably, the most damning of these artefacts is the impact of multiple blinking in which a single molecule may be localised multiple times. This can increase the density of points within a small region and therefore generate artefactual clusters. By applying pair correlation analysis to the statistical distribution of localisation errors, it is possible to derive the average size of clusters and filter out those which are far below the expected size. That said, it is impractical to attempt the analysis of experimental data without some degree of manual post-processing – at the very least, the resulting cluster distribution should be checked for outliers and extremities.

2.6.2 Cluster Analysis

Within a data set, a "cluster" can generally be defined as a subset of the data in which each point within that subset shares a suitable degree of similarity with some or all of the other points in that subset. In the case of spatial cluster analysis, this degree of similarity is often taken to be some degree of distance between points or the density of points within a specified range. However, since the definition of a cluster is inherently ambiguous, so too is the appropriate measure of quantifying it. As a result, a number of cluster analysis algorithms have been proposed, each exploiting the properties of the point clouds to derive some degree of segmentation between points and partition them into spatially aggregated groups. This partition is known as a "clustering" and is the goal of all cluster analysis algorithms.

While each algorithm differs in fundamental principles, parameters, hyperparameters, and runtime, they almost always incorporate at least one parameter (designed to absorb as much ambiguity and subjectivity as possible) which will dictate the distance or distances over which points may be considered similar. Additionally, they all produce a partitioning of the data points justified by the assumptions from which the algorithm was formed. As such, it is not always clear which, if any, method of cluster analysis is appropriate. In the following section, we will explore some of the most popular cluster analysis algorithms, outline their algorithmic processes, and describe when they might be useful.

Measurements of the spatial distribution of points which do not incorporate a precise partitioning of the points are known as spatially descriptive statistics. While these will not necessarily generate clusters from the data set, they are often paramount in informing the choice of parameters for which cluster analysis algorithms will be performed. The most notable of these spatially descriptive statistics are Ripley's functions, which will also be explored in the following section.

2.6.3 Choices of Algorithms

We begin our discussion with the most popular form of cluster analysis, known as density-based clustering. These methods stem from the principle that a cluster defined the region of an ROI in which the density of points is greater relative to the background (Nieves, Pike et al. 2023) (Figure 2.9). The most abundant density-based cluster analysis technique is known as Density-Based Spatial Clustering of Applications with Noise (DBSCAN). As this method is dependent on a measure of density, it requires both a length scale, ϵ, and a number of points, minPts. In practice, any points with at least minPts neighbours within the radius ϵ are denoted as "core" points, and any points which contain a core point within the radius ϵ (but are not core points themselves) are denoted as border points. A cluster is then defined as any group of neighbouring core and border points. While DBSCAN is simple, practical, and easy to employ, it is not always clear how to select the appropriate parameter values. In the field of SMLM and cluster analysis, over-parametrisation is common. This can be both a blessing and a curse as the correct choice of parameters could match manual human partitioning, but it also means that any possible clustering might

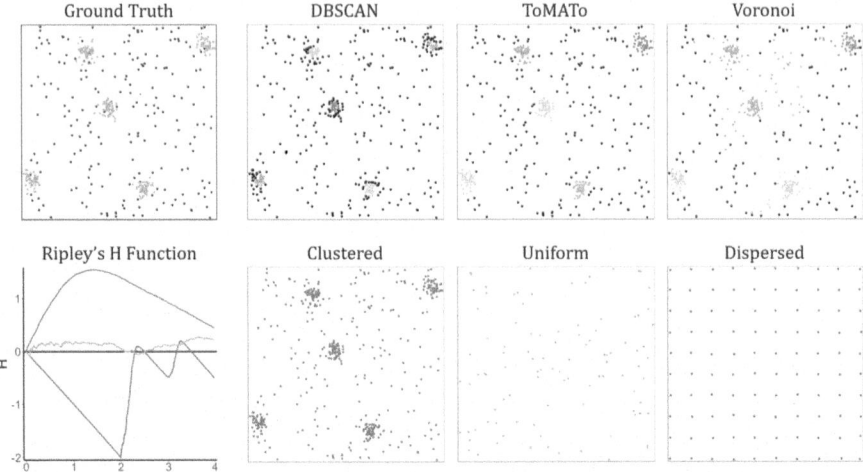

FIGURE 2.9 Examples of spatially descriptive statistics. A point cloud is shown alongside the results of DBSCAN, ToMATo and Voronoi clustering. In the bottom portion of the figure, Ripley's K-functions for clustered, uniform and dispersed data sets are given.

be equally appropriate. While other density-based methods exist, few are able to function with less than two informed parameters. These methods are particularly useful if the expected density of points and cluster size is known and will provide fast partitioning of the data.

An alternative approach known as Bayesian clustering employs statistical formulas regarding uncertainty to actively update the probability of each point belonging to a certain cluster (Rubin-Delanchy, Burn et al. 2015). This makes use of the renowned Bayes' formula which, in this case, describes the probability of the points adopting a certain partitioning conditional on the distribution of points. This process is carried out by first assigning each point a cluster allocation (randomly if necessary, although the more sensible the configuration, the faster convergence may be) known as a prior. Then, using some metric of the distribution of points, different allocations are systematically examined to determine an appropriate posterior distribution (that with the greatest probability of being the true cluster configuration), which is then selected as the new clustering.

A growing method of cluster analysis stems from the field of Topological Data Analysis (TDA). Topological cluster analysis methods are based on the principle of persistent homology, which is probed by evaluating the neighbours of each point over a range of possible radius values (e.g., by growing a disc around each point and connecting points when their discs touch). This produces the so-called Vietoris–Rips complex. Then, like other techniques in TDA, persistent homology checks for topological features of a data set – that is, the number of connected components and any holes which form in the connected components of the data set. As this examination is conducted over a range of radius values, it is possible to ascertain when the topological feature was formed (the birth scale) and when it was destroyed (the death scale) – the difference between the birth and the death scale is known as the persistence. Once this has been calculated for each topological feature, a persistence scale can be set below which features are disregarded. Birth scales can be plotted against death scales in a persistence diagram to inform the choice of this persistence scale. Studies have shown that clusters in a data set tend to have particularly high persistence, whereas outliers will not. Small variations in the position of points in a data set will have little overall effect on the persistence, making this method robust to noise. Additionally, this method detects clusters of arbitrary shape and size, which is not the case for more primitive algorithms. A method of TDA known as Topological Mode Analysis Tool (ToMATo) works back from persistent homology data to construct the clusters themselves and has been shown to outperform density-based algorithms.

As mentioned previously, Ripley's functions are spatially descriptive statistics, meaning they provide information about the dispersion of points and detect deviations in spatial homogeneity rather than giving a specific clustering. In this case, there are three functions, known as the K-, L-, and H-functions, with each building off of the last. The functions themselves determine if there is a statistically significant difference between the density of points within a specified range of distances and the expected density in a uniform distribution, averaged over each point in the ROI. The first of Ripley's functions is the K-function, which has functional form given by

$$K(r) = \frac{A}{n} \sum_{i=1}^{n} \sum_{j=1, j \neq i}^{n} H\left(\left|x_i - x_j\right| - r\right).$$

Here, A is the region of the ROI, n is the total number of points, x_i is the position of point i, r is the radius being probed for, and H is the Heaviside function, which evaluates to 1 if its input is positive and 0 otherwise. This translates to drawing a series of circles around each point, counting the number of points within each circle (not including the original point itself), then averaging over the density of the whole domain. This value is recorded for each value of r and can then be plotted. The K-function can be difficult to interpret as it has not been normalised. This normalisation is achieved through the L-function, defined simply as

$$L(r) = \sqrt{\frac{K(r)}{\pi}},$$

which acts as an intermediary to the H-function, given as

$$H(r) = L(r) - r.$$

This is arguably the most easily interpretable form of Ripley's function as it can highlight a variety of clustering behaviours over a range of search radii. Generally, a peak in the H-function at a certain radius suggests clustering is most prominent at that radius, which can inform the choice of search radius parameter for other clustering algorithms. Regions of the H-function close to 0 suggest that the distribution is approximately uniformly distributed across that range of radius values. Negative values of the H-function suggest homogeneity at the radii they are defined at.

For all of the techniques listed above, there are software packages available in many major programming languages, including Python and R.

In multicolour SMLM, multiple categories of molecules can be identified within the same image, yielding a marked point pattern depicting both the spatial coordinates of each localised molecule and that molecule's categorical value. This data is usually derived through the use of environmentally sensitive dyes, whose exact emission spectra depend on the bulk properties of the molecule they are attached to or the medium around them. This means that SMLM can detect discrete properties of each molecule and assign them a category, or that continuous data such as the degree of membrane order or pH can be determined for each molecule and then discretised to produce distinct categories. For this form of data, there are two properties we might be interested in: the degree of co-localisation, the extent to which distinct fluorescence channels overlap or correlate, and the degree of co-clustering, the extent to which molecules from distinct categories tend to cluster.

Since two molecules cannot occupy the same location, there will never be a precise overlap of their corresponding fluorescence channels, meaning that, theoretically, co-localisation cannot occur. However, due to localisation error, it is often more appropriate to view individual localisations as samples of probability densities, which can overlap for different colour channels if the molecules are closer than the spatial resolution will allow. In any case, it is imperative to correct for chromatic aberrations (e.g., via fluorospheres or multicolour beads) as this can offset the localisations of molecules imaged with particular colours of fluorophores. It helps to view the distinct channels from the image separately. In this instance, established methods for co-localisation from conventional microscopy will often suffice, although these results can be impacted by background noise and stochastic blinking. More recently, spatially descriptive statistics have been adapted to incorporate multiple channels. Additionally, techniques for determining the degree of co-clustering have been adapted from traditional methods of analysing spatial clustering (Simoncelli, Griffié et al. 2020).

2.7 BIOLOGICAL EXAMPLES

2.7.1 Neurons

Signal transduction across the axon membrane relies almost entirely on the abundance of protein channels, and as a result, the axon is typically densely packed with proteins. This makes them impractical to image with conventional microscopy. However, 3D STORM has been used to highlight the existence of actin rings wrapped around the circumference of the axon, with rings separated by sub-diffraction scales (Xu, Zhong et al. 2012). In neuronal cells, PALM has been used to image F-actin molecules in the cellular protrusions known as dendrites. The actin cytoskeleton within the dendritic spine is thought to bring about perturbations to the geometry of the spine. By using PALM, it was revealed that the dynamics of F-actin are highly heterogeneous, which implies that the actin filaments in the dendritic spine are short. Some data analysis methods have been devised to segment point clouds and identify biological structures within them. This is particularly useful in this biological context as it allows for probing the morphology of these structures. In the case of neurons, this can allow for analysis of the orientation, length, and curvature of dendritic spines in neurons.

2.7.2 Immune Cells

The immune system functions almost entirely through the identification of antigens and the usage of receptors in identifying them. Arguably, the most prominent cell involved in this process is the T cell, whose main role is to identify harmful antigens. Studies have found that the membrane proteins T cell receptor (TCR) and linker for activation of T cells (LAT) form distinct clusters prior to T cell activation but then merge subsequently (Williamson, Owen et al. 2011). It is hypothesised that clustering of TCR at the nanoscale plays a role in antigen recognition and signal initiation. Other studies dispute this claim, but suggest that the distribution of TCRs on the plasma membrane is actually optimised for efficient antigen recognition in the first phase of T cell activation. In any case, SMLM is paramount to furthering research into receptor distribution on the plasma membrane of immune cells.

2.7.3 Cancer Cells

Previously, dSTORM has been used to probe clustering of tumour necrosis factor receptor (Karathanasis, Medler et al. 2020). Other studies have employed SMLM techniques to image metastasis-specific microRNA molecules in subcellular sites on cancer cells. Quantitative analysis revealed that these molecules were heterogeneously distributed and the stoichiometry differed between weakly metastatic cells and highly metastatic cells, potentially revealing a causal link to cancer.

REFERENCES

Betzig, E., G. H. Patterson, R. Sougrat, O. W. Lindwasser, S. Olenych, J. S. Bonifacino, M. W. Davidson, J. Lippincott-Schwartz and H. F. Hess (2006). "Imaging intracellular fluorescent proteins at nanometer resolution." *Science* **313**(5793): 1642–1645.

Bohrer, C. H., X. Yang, S. Thakur, X. Weng, B. Tenner, R. McQuillen, B. Ross, M. Wooten, X. Chen, J. Zhang, E. Roberts, M. Lakadamyali and J. Xiao (2021). "A pairwise distance distribution correction (DDC) algorithm to eliminate blinking-caused artifacts in SMLM." *Nature Methods* **18**(6): 669–677.

Culley, S., D. Albrecht, C. Jacobs, P. M. Pereira, C. Leterrier, J. Mercer and R. Henriques (2018). "Quantitative mapping and minimization of super-resolution optical imaging artifacts." *Nature Methods* **15**(4): 263–266.

Dempsey, G. T., J. C. Vaughan, K. H. Chen, M. Bates and X. Zhuang (2011). "Evaluation of fluorophores for optimal performance in localization-based super-resolution imaging." *Nature Methods* **8**(12): 1027–1036.

Heilemann, M., S. van de Linde, M. Schüttpelz, R. Kasper, B. Seefeldt, A. Mukherjee, P. Tinnefeld and M. Sauer (2008). "Subdiffraction-resolution fluorescence imaging with conventional fluorescent probes." *Angewandte Chemie International Edition in English* **47**(33): 6172–6176.

Huang, B., W. Wang, M. Bates and X. Zhuang (2008). "Three-dimensional super-resolution imaging by stochastic optical reconstruction microscopy." *Science* **319**(5864): 810–813.

Juette, M. F., T. J. Gould, M. D. Lessard, M. J. Mlodzianoski, B. S. Nagpure, B. T. Bennett, S. T. Hess and J. Bewersdorf (2008). "Three-dimensional sub-100 nm resolution fluorescence microscopy of thick samples." *Nature Methods* **5**(6): 527–529.

Karathanasis, C., J. Medler, F. Fricke, S. Smith, S. Malkusch, D. Widera, S. Fulda, H. Wajant, S. J. L. van Wijk, I. Dikic and M. Heilemann (2020). "Single-molecule imaging reveals the oligomeric state of functional TNFα-induced plasma membrane TNFR1 clusters in cells." *Science Signaling* **13**(614): 1–10.

Lelek, M., M. T. Gyparaki, G. Beliu, F. Schueder, J. Griffié, S. Manley, R. Jungmann, M. Sauer, M. Lakadamyali and C. Zimmer (2021). "Single-molecule localization microscopy." *Nature Reviews Methods Primers* **1**: 1–27.

Manley, S., J. M. Gillette, G. H. Patterson, H. Shroff, H. F. Hess, E. Betzig and J. Lippincott-Schwartz (2008). "High-density mapping of single-molecule trajectories with photoactivated localization microscopy." *Nature Methods* **5**(2): 155–157.

Marsh, R. J., I. Costello, M.-A. Gorey, D. Ma, F. Huang, M. Gautel, M. Parsons and S. Cox (2021). "Sub-diffraction error mapping for localisation microscopy images." *Nature Communications* **12**(1): 5611.

Nieves, D. J., J. A. Pike, F. Levet, D. J. Williamson, M. Baragilly, S. Oloketuyi, A. de Marco, J. Griffié, D. Sage, E. A. K. Cohen, J.-B. Sibarita, M. Heilemann and D. M. Owen (2023). "A framework for evaluating the performance of SMLM cluster analysis algorithms." *Nature Methods* **20**(2): 259–267.

Ovesný, M., P. Křížek, J. Borkovec, Z. Švindrych and G. M. Hagen (2014). "ThunderSTORM: a comprehensive ImageJ plug-in for PALM and STORM data analysis and super-resolution imaging." *Bioinformatics* **30**(16): 2389–2390.

Rubin-Delanchy, P., G. L. Burn, J. Griffié, D. J. Williamson, N. A. Heard, A. P. Cope and D. M. Owen (2015). "Bayesian cluster identification in single-molecule localization microscopy data." *Nature Methods* **12**(11): 1072–1076.

Rust, M. J., M. Bates and X. Zhuang (2006). "Sub-diffraction-limit imaging by stochastic optical reconstruction microscopy (STORM)." *Nature Methods* **3**(10): 793–796.

Schnitzbauer, J., M. T. Strauss, T. Schlichthaerle, F. Schueder and R. Jungmann (2017). "Super-resolution microscopy with DNA-PAINT." *Nature Protocols* **12**(6): 1198–1228.

Simoncelli, S., J. Griffié, D. J. Williamson, J. Bibby, C. Bray, R. Zamoyska, A. P. Cope and D. M. Owen (2020). "Multi-color molecular visualization of signaling proteins reveals how C-terminal Src kinase nanoclusters regulate T cell receptor activation." *Cell Reports* **33**(12): 108523.

Thompson, R. E., D. R. Larson and W. W. Webb (2002). "Precise nanometer local-
    ization analysis for individual fluorescent probes." *Biophysical Journal* **82**(5):
    2775–2783.

Williamson, D. J., D. M. Owen, J. Rossy, A. Magenau, M. Wehrmann, J. J. Gooding
    and K. Gaus (2011). "Pre-existing clusters of the adaptor Lat do not partici-
    pate in early T cell signaling events." *Nature Immunology* **12**(7): 655–662.

Xu, K., G. Zhong and X. Zhuang (2012). "Actin, spectrin, and associated proteins
    form a periodic cytoskeletal structure in axons." *Science (New York, N.Y.)*
    339: 452–456.

CHAPTER 3

Measuring Diffusion of Membrane Biomolecules Using Fluorescence Microscopy

3.1 THE IMPORTANCE OF DIFFUSION

3.1.1 Lipid Organisation and Effects on Molecular Diffusion

Approximately 50% of the plasma membrane consists of phospholipids whose chemical structure can be diverse, i.e., their fatty acid tails can be saturated or unsaturated, and their head groups can bear functional chemical groups (Harayama and Riezman 2018). An important property of the lipid bilayer is that it acts like a two-dimensional fluid, thus individual molecules (both lipids and proteins) can move freely within it. It is now clear that the distribution of these lipids in the bilayer is not homogeneous or random, and discrete lipid domains can be formed therein, i.e., liquid ordered domains (rich in cholesterol and saturated lipids) and liquid disordered domains (abundance of unsaturated lipids; Bacia, Scherfeld et al. 2004; Harayama and Riezman 2018; Kinoshita, Suzuki et al. 2018).

Lipid domains are thought to play a critical role in the regulation of biomolecule diffusion within the membrane. The structure of the lipids within these domains can modify the fluidity of the membrane, e.g., unsaturated lipids, due to their kinked fatty acid tails, do not pack as tightly as saturated lipids, thus decreasing their packing (Harayama and Riezman 2018). Generally, less tightly packed membrane is more fluid, thus this would facilitate the increased diffusion of proteins and lipids within these regions. Furthermore, tighter packing of lipids in ordered domains can also contribute to the slowing and trapping of molecules in the membrane, which can facilitate protein oligomerisation (Beckers, Urbancic et al. 2020).

3.1.2 Initiation and Inhibition of Cell Signalling

Cell signalling is a process whereby cells respond to their local environment, i.e., molecules outside of the cell such as hormones, neurotransmitters, and growth factors, through signalling molecules within the plasma membrane (Schlessinger 2000; Kholodenko 2006; Gurevich and Gurevich 2019). Generally, signalling is initiated by the binding of an extracellular ligand to a membrane-embedded receptor protein, inducing a conformational change, and leading to activation of the receptor's catalytic functions (e.g., receptor tyrosine kinases; RTKs) or that of their signalling partners (G-protein coupled receptors; GPCRs; Schlessinger 2000; Gurevich and Gurevich 2019). Critically, in both cases, the assembly of receptor:ligand complexes and subsequent downstream interactions are all facilitated by diffusion. RTKs often require dimerisation for full activation, which is mediated by the lateral movement of RTK monomers in the plasma membrane as well as diffusion of the ligand to the site of the receptor (Schlessinger 2000). Similarly, the association and dissociation of the G-proteins with the hormone receptor is required for function (Gurevich and Gurevich 2019).

Similarly, the diffusion of proteins inhibitory to cell signalling is also a critical component of signalling regulation. One such example is the lateral movement of phosphatases in the plasma membrane (Brown and Kholodenko 1999). Briefly, phosphatases remove phosphate groups from proteins, and this is of relevance to RTKs, which become phosphorylated upon activation, as these proteins can attenuate their signalling (Ostman and Bohmer 2001). Critically, proximity is required for phosphatase action on active receptors; thus, diffusion of phosphatases in the membrane allows signalling to be regulated. This has been demonstrated most notably in the context of T-cell receptor (TCR) signalling regulation by the phosphatase CD45 (Cordoba, Choudhuri et al. 2013; Razvag, Neve-Oz et al. 2018). Here, the lateral movement of CD45 in the membrane is key to its subsequent exclusion from the immune synapse, owing to its large ectodomain, upon T-cell activation (Irles, Symons et al. 2003; Cordoba, Choudhuri et al. 2013; Razvag, Neve-Oz et al. 2018). This means that the TCR, specifically the associated CD3 complex, can remain phosphorylated when bound to antigen when a tight contact with the antigen-presenting cell is formed. This mechanism is thought to underpin a large part of T-cell sensitivity to low levels of antigen (Furlan, Minowa et al. 2014; Razvag, Neve-Oz et al. 2018; Pettmann, Huhn et al. 2021).

3.1.3 Cell Adhesion and Migration

One critical function of the plasma membrane is to maintain the structural integrity of the cell and regulate the morphology of cells to support adhesion and migration in response to chemical cues on their substrate (Buckley, Rainger et al. 1998; Cavalcanti-Adam, Micoulet et al. 2006; Cavalcanti-Adam, Volberg et al. 2007; Bachmann, Kukkurainen et al. 2019; Shannon, Pineau et al. 2019). For this to occur, strict control over the movement, diffusion, and distribution of adhesion receptors is paramount (Cavalcanti-Adam, Micoulet et al. 2006; Cavalcanti-Adam, Volberg et al. 2007). For example, in the formation of nascent cell adhesions, integrin clusters were found to be of consistent size regardless of the substrate and that these clusters must recruit unliganded integrins, with interaction between integrins or integrin-adhesion proteins stabilising new molecules in the clusters and therefore increasing the size of the adhesion. Individual liganded integrins diffuse in the plasma membrane but are rapidly removed via retrograde flow of linked actin cytoskeleton (Felsenfeld, Choquet et al. 1996). Furthermore, affinity of integrins for adhesive ligands is weaker prior to the leading edge of the cell membrane, thus enabling integrins to diffuse to the initial nascent adhesions and assemble. For example, α^4 integrin adhesion to vascular cell adhesion molecule (VCAM)-1, whereby mutation of the intracellular domain of α^4 integrin causes defective adhesion, although VCAM-1 binding was not affected (Yauch, Felsenfeld et al. 1997). It was observed that the lateral mobility of α^4 integrin was impaired, thus maturation of focal adhesions was hindered (Yauch, Felsenfeld et al. 1997).

3.2 TECHNIQUES FOR MEASURING BIOMOLECULE DIFFUSION

3.2.1 Fluorescence Recovery after Photobleaching (FRAP)

Given that fluorophores and fluorescent proteins are prone to bleaching, it is possible to exploit this to measure ensemble movement and diffusive properties of labelled molecules. Further, depending on the matrix, or closed environment, observed, one can infer information of the structure of diffusive gradients therein, e.g., regions of higher and lower molecular diffusion. Thus, Axelrod et al. developed a technique to exploit these phenomena and provide information on the average dynamics of fluorescently labelled molecules, termed fluorescence recovery after photobleaching (FRAP; Axelrod, Koppel et al. 1976). To make such measurements, first a defined region of the closed environment is excited with high excitation power (Figure 3.1). The aim here is to fully bleach all the fluorescently

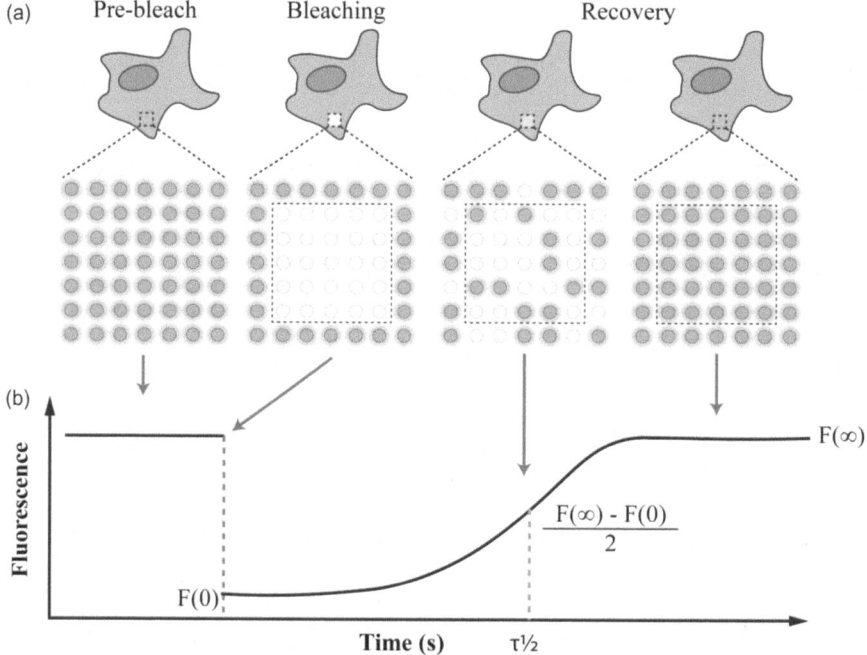

FIGURE 3.1   (a) Prior to bleaching, the membrane is completely covered by fluorescent molecules, and a region is selected to perform the FRAP measurement. The selected region (dotted black line) is then illuminated with a high laser excitation power, sufficient to bleach all the molecules within the region. Diffusion of fluorescent molecules from the non-bleach areas of the cell into the bleached region leads to recovery of the fluorescent signal within the bleached region. Eventually, the fluorescence is completely recovered within the region in mobile systems. (b) Example FRAP measurement curve for the scenario in (a). Upon bleach, the measured fluorescence intensity drops to a minimum value ($F(0)$), and over time, the fluorescence intensity returns to the steady state ($F(\infty)$). FRAP curves can be simply quantified by calculating the half time of recovery ($\tau\frac{1}{2}$), which is the time required for a bleached area to recover half the steady state fluorescence.

labelled molecules within this region, rendering these molecules undetectable in subsequent images (Axelrod, Koppel et al. 1976; Lippincott-Schwartz, Snapp et al. 2018). Following this bleaching step, the region is imaged over a defined time window on the order of seconds to minutes. If the molecules are freely diffusive, the molecules within the bleached region are free to leave this area, and further, the molecules that were outside of the bleaching area can diffuse in. Only the molecules that were outside of the bleached area can now be detected within this region (*recovery*) (Axelrod, Koppel et al. 1976; Lippincott-Schwartz, Snapp et al. 2018). Imaging this region over time with lower excitation power, to avoid/limit further photobleaching, allows the measurement of the recovery of fluorescence signal within this region over time (Figure 3.1). The rate at which fluorescence signal is replenished within the bleached region, as opposed to a non-bleached control region, is used to extract the ensemble diffusion coefficient for the molecules within the closed environment (Axelrod, Koppel et al. 1976; Lippincott-Schwartz, Snapp et al. 2018). Depending on the sample and the size of the bleaching region, it may be possible to do several measurements on the same cell and thus start to access differences in the structure of the environment in terms of molecular diffusion, e.g., lipids, membrane, and ECM (Goehring, Chowdhury et al. 2010; Mudumbi, Schirmer et al. 2016; Sun, Marcello et al. 2016; Oshima, Nakashima et al. 2019).

3.2.2 Single Particle and Single Molecule Tracking

Generally, tracking requires the acquisition of sequential images of a sample with sparse labelled or tagged molecules and allows the motion of individual molecules to be observed, given the rate of image acquisition is fast enough. The position of each molecule can be determined within each image within the sequence, and once linked, a time-course generated of the diffusion behaviour of a single molecule (known as a trajectory, Figure 3.2; Saxton and Jacobson 1997). This was demonstrated in pioneering work investigating the lateral diffusion of individual proteins (E-cadherin and Transferrin) in the membrane (Kusumi, Sako et al. 1993) and allowed the identification of protein confinements in the plasma membrane, giving the first insights in protein–cytoskeletal corralling interactions (Sako and Kusumi 1994). Here, the proteins were labelled with large and non-fluorescent latex beads and exploited light scattering as the detection signal, and this can largely be described as part of the larger field of single particle tracking (SPT). One drawback of SPT methods is linked

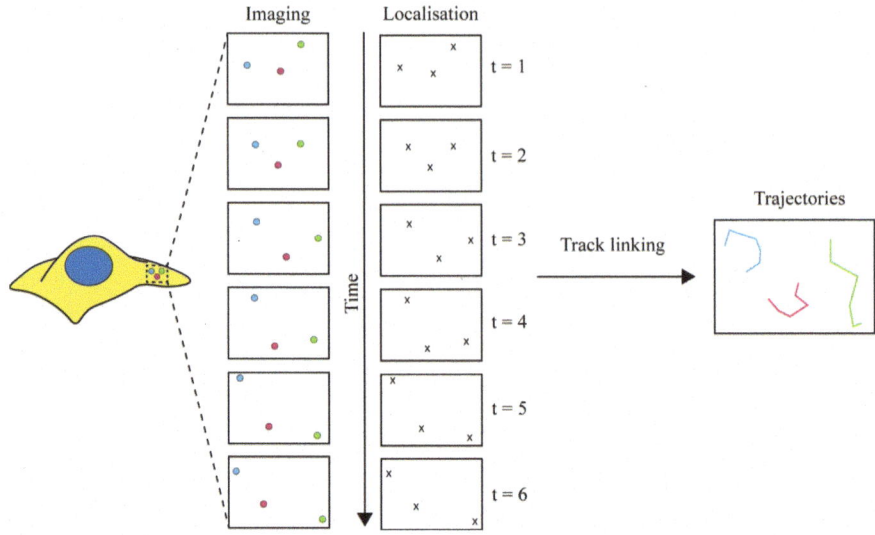

FIGURE 3.2   Sequential images of single fluorescent diffusing molecules (cyan, magenta, green) at the cell surface are acquired ($t=1$ to $t=6$). For each image frame, the centre of the molecules (X) at that timepoint (t) is determined by fitting. Once all the molecules have been localised, they can be linked into a trajectory depending on the linking parameters.

to the labelling itself. For example, the need to generate a scattering signal greater than the cell background meant that the labels had a diameter of 40 nm or more (Sako and Kusumi 1994), which could constrain the molecules diffusion due to steric interactions of the label with the environment. Furthermore, controlling the number of these labels per molecule was also a challenge and these methods could lead to crosslinking and aggregation of molecules (Iyer, Michalet et al. 2010). Thus, with the development of genetically encoded fluorescent proteins (*e.g.*, GFP; Heim, Prasher et al. 1994; Tsien 1998), exploiting fluorescence as a signal for single molecule tracking (SMT) gathered increasing interest. Additionally, because fluorescent proteins are much smaller than their particulate counterparts, they have less of a steric effect on the molecules they are attached to. Therefore, there has been significant development and usage of fluorescent labels for this purpose, e.g., organic dyes (Fernandez-Suarez and Ting 2008; Benke, Olivier et al. 2012) and photoactivation (Patterson and Lippincott-Schwartz 2002; Wiedenmann, Ivanchenko et al. 2004), and now SMT is the most prevalent form of tracking used to observe molecular diffusion within the membrane (Fernandez-Suarez and Ting 2008; Manley, Gillette et al. 2010).

3.2.3 Fluorescence Correlation Spectroscopy (FCS)

Fluorescent molecules diffusing through a detection volume, for example, the diffraction limited point spread function (PSF) of a confocal microscope, allow the residence time of those molecules to be captured (Magde, Elson et al. 1972). Recording a time trace of fluorescent molecules diffusing through the detection volume yields a fluorescence intensity trace with the number of peaks and their width related to the concentration and diffusion speed of the molecules observed. To extract an average diffusion coefficient and concentration from these fluorescence time traces, the traces are correlated in time with themselves, known as autocorrelation (Magde, Elson et al. 1972). The resulting autocorrelation curve starts at a maximal value and decays as the fluorescent trace is shifted by a defined time interval (typically tens of μs) and thus become less correlated with itself. The rate of decay of the autocorrelation function is determined by the width of the peaks within the trace, i.e., the length of time the molecule spends in the detection volume, whereas the maximal value of the function is determined by the concentration of molecules (Figure 3.3) (Magde, Elson et al. 1972). The autocorrelation curve can be fitted with a model that describes

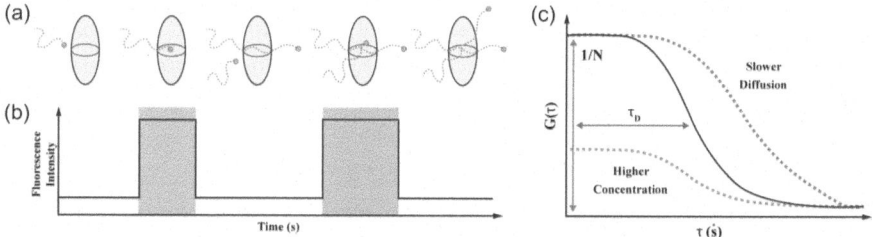

FIGURE 3.3   (a) Single fluorescent molecules diffuse through the confocal measurement volume generated by the point spread function of the microscope. (b) Measuring the fluorescence intensity of the molecules passing through the volume allows a trace to be acquired containing fluorescence fluctuations owing to diffusion (blue and orange). (c) Autocorrelation of the trace at time t with all points corresponding to t plus the delay time (τ) allows an autocorrelation curve to be determined (black). This curve gives information in the concentration (from the inverse of $1/N$, green arrow) and the diffusion coefficient of the molecules (calculated using τD, brown arrow). The shape and size of the curve is sensitive to the concentration and diffusion speeds within the sample (cyan and magenta dashed lines).

the diffusion of fluorescent molecules through a detection volume, and thus, the average diffusion coefficient of the molecules at that point within the sample can be determined. This method is generally known as auto-correlation spectroscopy, and in the context of fluorescence microscopy, fluorescence correlation spectroscopy (FCS; Magde, Elson et al. 1972).

3.2.4 Comparison and Limitations of FCS, FRAP, and SMT

In general, the above techniques can be split broadly into two groups: those that give ensemble diffusion measurements (i.e., FRAP and FCS) and those that allow single molecule diffusion coefficients to be calculated (SMT). Both FRAP and FCS are diffraction limited measurements, limiting the smallest area that can be observed by these techniques (>~250 nm). However, it should be noted that there are subtleties in terms of measurement scale that set them apart from each other. FCS can largely only make single point measurements; thus, coverage of the plasma membrane and measurable concentration ranges will ultimately be limited (Blom, Kastrup et al. 2006; Hell 2009), while FRAP can observe larger portions of the membrane on the micron scale (Lippincott-Schwartz, Snapp et al. 2018). But the drawback still inherent in these techniques is that regardless of the scale, heterogeneity within these observed regions will be lost by averaging.

In the case of SPT, given it observes isolated molecules, sub-diffraction information on the diffusion dynamics of molecules can be obtained. This means that it is possible to measure the rare and heterogeneous diffusive behaviour present within a population of molecules, e.g., describe whether their diffusion is Brownian or anomalous or measure molecule confinements (Ritchie, Shan et al. 2005). However, given that SPT observes a subset of molecules, normally many trajectories (in the range of several thousand) are required to make statistically robust conclusions about rare diffusion events within the sample (Walder, Kastantin et al. 2012). Therefore, this often requires many measurements of different cells and several samples.

3.3 USING FCS TO MEASURE DIFFUSION OF BIOMOLECULES

3.3.1 Hardware

Generally, FCS is performed using confocal laser scanning microscopes (CLSM) as they can most easily generate the illumination conditions required for the technique, i.e., small detection volume, and are easily amenable for live-cell imaging (Elliott 2020; Yu, Lei et al. 2021), although it should be noted that some wide field implementations have recently been developed (Buchholz, Krieger et al. 2018). The excitation can be either single photon or two photons, depending on the experimental requirements, with two-photon excitation reducing the size of the detection volume, thus sampling a smaller region in the sample (Berland, So et al. 1995; Schwille, Haupts et al. 1999; Yu, Lei et al. 2021). In a CLSM, the detection volume is essentially the point spread function of the system, i.e., a convolution of the diffraction limited focus of the excitation laser illumination (using an objective with NA > 1; Elliott 2020) and the size of the confocal pinhole for collection, and the size of this volume is referred to as V_0. Critical to accurate FCS measurements is accurate knowledge of the value for V_0 and is typically determined by measurement of a z-scan of a sub-diffraction (<100 nm) fluorescent bead through the volume (Ruttinger, Buschmann et al. 2008).

Fluorescence detection for FCS requires on the sensitive capture of the photons coming from the volume; hence, photomultiplier tubes (PMTs) or single-photon avalanche photodiodes (SPADs) are the preferred choices (Yu, Lei et al. 2021). The only additional tool(s) required for FCS acquisition would be a way to correlate the recorded fluorescence signal, which traditionally may have used a hardware-based correlator. Now the autocorrelation is largely done by software packages, which are either available online or integrated into the acquisition software of the microscope; thus, the measured traces are stored on a computer and can be directly accessed and analysed (Yu, Lei et al. 2021).

3.3.2 Fluorescent Labels for FCS

In terms of fluorescent labelling of molecules for FCS, the technique is reasonably flexible. Any commonly used fluorescent labelling system, i.e., expressed fluorescent proteins or organic dyes, will likely be suitable for FCS measurements (Schwille 2001; Haustein and Schwille 2007). However, for expression systems, it is advisable to avoid high overexpression of the protein of interest as this will compromise the data quality, leading to saturated traces and lack of fluctuations to analyse (Kim, Heinze et al. 2007). Importantly, the fluorescent signal for a single emitter needs to be high enough for detection, i.e., high photon count per molecule. This is particularly critical for situations where the molecular concentration is low or where there may be background fluorescence within the samples (e.g., within cells; Kim, Heinze et al. 2007).

However, there are some artefacts owing to the stability of the fluorescence emitted from the label that should be avoided to maintain the integrity of the FCS spectra acquired. For example, high photostability is particularly desired if the diffusion kinetics of the molecule being observed are not known, as slow-moving molecules, which may be excited for long times in the volume, are prone to bleaching (Hodges, Kafle et al. 2018). This will lead to an artificially reduced transit time of the molecule through the volume and an overrepresentation of fast diffusion in the sample (Hodges, Kafle et al. 2018). This is also true for quantum dot labels; although they exhibit the high photon counts desired for an FCS label, they have intermittent fluorescent blinking kinetics which ultimately will contribute to the measured FCS spectra, again decreasing the apparent residence time within the volume and speeding up the measure diffusion coefficient (Heuff, Swift et al. 2007).

3.3.3 FCS Data and Processing

The data extracted from FCS experiments is normally in the form of a time trace where fluorescent signal fluctuations corresponding to fluorophores diffusing through the confocal volume were recorded (Yu, Lei et al. 2021). The key aspects of FCS data analysis are processing of the raw traces and fitting the data with theoretical model to extract relevant information such as diffusion coefficients, molecular concentrations, and interactions (Yu, Lei et al. 2021).

The FCS data analysis pipeline typically follows the following steps:

1. **Pre-processing of raw traces:** It might be beneficial to pre-process the raw traces acquired from an FCS measurement to remove unwanted signal contributions which will detrimentally affect the calculation of molecule concentrations. For example, large aggregates of protein within a sample will generate high intensity peaks within the raw spectra, several times that of single molecules.

2. **Correlation of fluorescence intensity traces:** The single molecule fluorescence traces collected from the microscope are correlated in time. This can either be correlation with itself (autocorrelation) or with another contemporaneous signal, if acquired (cross-correlation). The resulting autocorrelation curve is plotted as a function of the

lag time (τ), which is the time (in seconds) for a single measurement interval (Figure 3.3).

3. **Model selection and fitting:** There are several different models that can be used to fit FCS data depending on the nature of the information being sought (Yu, Lei et al. 2021), e.g., a simple model might be used to study the diffusion coefficient of a single species in solution (diffusion-limited Gaussian model):

- **Gaussian model:** This assumes that the fluorescence intensity detected is proportional to the number of molecules within a Gaussian observation volume and the molecular movement is owing to 3D diffusion:

$$G(\tau) = \frac{1}{\left(1 + (\tau/\tau_D)\right)\left(1 + a^{-2}(\tau/\tau_D)\right)^{1/2}}$$

where $a = w_z/w_{xy}$ is the ratio of the axial and radial radii of the volume and τ_D is the average residence time of molecules in the detection volume. This equation is used to fit the autocorrelation data and extract typically two parameters: $G(0)$, which is the intercept at the y axis at the first lag time, and τ_D (Figure 3.3).

Several models exist for various biological scenarios which may be present within the experiments, e.g., interaction of two labelled proteins, and may also be applied accordingly. Once selected, the data are fitted with the model, and the quality of the fit can be assessed using various metrics, such as the goodness-of-fit, residuals, and correlation coefficients (Yu, Lei et al. 2021).

4. **Interpretation:** Values of the parameters obtained from fitting of the model can now be used to determine the concentration of molecules C calculated from $G(0)$:

$$G(0) = \frac{1}{N} = \frac{1}{\pi^{3/2}\omega_{xy}^2\omega_z C}$$

and the diffusion coefficient is given by

$$D = \frac{\omega_{xy}^2}{4\tau_D}$$

3.4 USING SMT TO MEASURE DIFFUSION OF BIOMOLECULES

3.4.1 Hardware

Often SMT is acquired using widefield microscopy techniques; however, it should be noted that some implementations use mechanical tracking and triangulation of single emitters within the sample (e.g., photothermal tracking (Duchesne, Octeau et al. 2012) and MINFLUX (Balzarotti, Eilers et al. 2017)). Widefield approaches appear to be the preferred option as to generate the trajectory numbers (tens of thousands) and robust statistics for diffusion analysis, observing several molecules simultaneously, although sparsely, greatly speeds up this process (Wieser and Schutz 2008). Thus, techniques that can acquire a large field of view (FOV) but maintain low emitter density and contributions from out of plane molecules is desired. For this reason, total internal reflection fluorescence (TIRF) or single plane illumination microscopy (SPIM) are preferred (Wieser and Schutz 2008). Generally, most standard laser excitation corresponding to the wavelength of the probe is suitable, but for experiments that require photoactivation, both the excitation wavelength of the probe and the photoactivation wavelength must be catered to.

Furthermore, these techniques routinely use high-speed camera detection for fluorescence sensing of molecules in real time, which is a critical component necessary for SMT (Wieser and Schutz 2008). Typically, EM-CCD or sCMOS cameras are used for this, as they generally have high framerates (~30 and ~100 fps, respectively), but also have high sensitivity and dynamic range (Manzo and Garcia-Parajo 2015). The remaining component required for SMT is a computer for both control of the microscope and camera, but also with large storage for the images stacks that may be acquired over several cells (in the Gb range).

3.4.2 Fluorescent Labels for SMT – Conjugation and Properties

As earlier discussed, labels for SMT tracking tend to be small in comparison to particle labels, and thus have little steric impact on the labelled molecules diffusion. Conventionally, fluorescent proteins, which can be expressed in fusion with the protein, or fluorescent dyes, which can be protein conjugated via several biochemical methods (e.g., click chemistry and self-labelling enzyme tags; Juillerat, Gronemeyer et al. 2003; Los, Encell et al. 2008; Nikic, Kang et al. 2015), have been used extensively for SMT. Fluorescent proteins have been applied extensively to observe the diffusion and interaction of proteins in living cells; however, such labels bleach under continuous excitation. Thus, photobleaching dramatically decreases the trajectory time length of a single molecule experiment to a maximum of ~10 s (often ~ 1 s) in an oxygen containing atmosphere. This means that the number of tracks required for meaningful analysis might be increased for such labelling. In the case of fluorescent dyes, they are much more photostable, and trajectory length can therefore be increased. However, being able to label a molecule with a single dye is still somewhat of a challenge (Nieves and Baker 2021), but several methods are now available for proteins (Juillerat, Gronemeyer et al. 2003; Los, Encell et al. 2008; Nikic, Kang et al. 2015).

For both label types, where expression of a fluorescent tag or enzyme tag is required, a concern for SMT is still control of the density of emitters that will be observed in the sample. Overexpression will often lead to fluorescent molecule densities too high for SMT. This drawback has been somewhat overcome by the development of photoactivatable (PA) fluorophores for single molecule imaging (Ando, Hama et al. 2002; Patterson and Lippincott-Schwartz 2002; Betzig, Patterson et al. 2006). Briefly, PA-fluorophores are weak or non-fluorescent until they are activated by an appropriate wavelength of light (typically around 400 nm), making them fluorescent. This means that the density of emitters can be controlled in real time on the microscope by tuning this photoactivation intensity. In turn, the expression level in terms of the SMT poses less of an issue for tuning emitter density and background, and many more tracks can be acquired from the same cell with repeated photoactivation. Furthermore, this allows tracking to be achieved in systems at higher concentrations possible for standard fluorescent probes. For example, both photoactivatable proteins (e.g., PA-mcherry, mEos3) and fluorophores (Janelia Fluors; PA-JF549 and PA-JF646 (Grimm, English et al. 2016)) have been developed and demonstrated for SMT in live cells (Manley, Gillette et al. 2010; Grimm, Muthusamy et al. 2017; Hilzenrat, Pandzic et al. 2020).

3.4.3 SMT Data and Processing

Commonly, data from SMT experiments are a stack of sequential images taken of the sample. As noted earlier, scanning and single trajectory triangulation-based methods, e.g., MINFLUX, data will likely already be in the form of recorded trajectories. For SMT image stacks, the data must be processed to generate the trajectories, which can then be used to make measurements on the type and speed of diffusion within the sample. In both cases, it is important to consider the following data properties before proceeding:

1. **Signal-to-noise (S/N) ratio:** Briefly, S/N ratio is the measure of the signal intensity divided by the variation in the background noise; thus, for data with a poor S/N ratio (<5), tracking will be problematic due to the single emitter being hard to detect over the background (Cheezum, Walker et al. 2001).

2. **Acquisition frame rate:** The frame rate is normally measured in number of images acquired per second. This must be tuned to be fast enough to detect molecule movements giving good coverage of the potential diffusion speeds but maintain high enough S/N ratio to allow localisation.

3. **Emitter density:** Low emitter density is critical for being able to localise single emitters. Increasing density (>1 emitter/μm^2) increases the difficulty of detecting single emitters and can lead to erroneous or localisation of molecules. Further, the correct linking of the same molecule between frames into a single trajectory.

Once the data is judged to be of sufficient quality for SMT analysis, it can be used to perform tracking analysis typically following the below pipeline:

1. **Single molecule localisation:** Firstly, the image data may be pre-processed using an image filter to reduce the contributions of camera noise in the image and enhance intensity peaks from single molecules within the image (Martens, Turkowyd et al. 2021). The position of these peaks within each image is then determined by peak fitting algorithms (Wieser and Schutz 2008). The most used localisation methods for SMT are least squares (LS) fitting and maximum likelihood estimation (MLE), and both normally assume a Gaussian model of the point spread function (Shen, Tauzin et al. 2017). LS requires little knowledge about the camera noise and experimental setup, and it can thus adapt to non-Gaussian PSFs, but lacks precision in low S/R settings. MLE on the other hand, rather than fitting the image data, attempts to determine the parameters that generated the current data, thus detailed knowledge of the experimental conditions and PSF are required, but it tends to be more precise than LS (Shen, Tauzin et al. 2017).

2. **Trajectory generation:** From here the x- and y-coordinates of the molecule centres are linked in time, i.e., frame to frame, to give single molecule trajectories depending on the following general parameters:

- **Capture radius:** This value is the radius of the area of search for another centroid in the next image of the sequence, i.e., a new centroid must be found within this radius from the last centroid for the points to be linked. If there is no point identified, the track will be terminated. This process can be determined by hard thresholds on the radius (Saxton 2008) or using models for the most likely distribution of the centroids, including Kalman filtering and Bayesian methods (Shen, Tauzin et al. 2017).

- **Minimal track length (frames):** Candidate trajectories can then be filtered depending on their length, and it is normally determined by the number of steps (which will equal a frame number) within the trajectory. Normally, trajectories that fall below this threshold are discarded, as they will lack the minimal information required to accurately extract diffusion behaviour (Wieser and Schutz 2008).

    Only trajectories passing these criteria are retained and deemed suitable for further analysis.

3. **Trajectory analysis:** Once the particle trajectories have been obtained, the next step is to extract key parameters which will inform about the diffusion behaviour of the molecules, i.e., diffusion coefficients and residence times. Currently, there are several approaches that can be used, depending on the information sought; however, commonly used methods are mean-squared displacement (MSD) analysis (Duchesne, Octeau et al. 2012) and confinement analysis (Simson, Sheets et al. 1995; Hilzenrat, Pandzic et al. 2020).

- **MSD analysis:** Generally, MSD is a measure for the deviation of a molecule from a reference position, normally the start of the trajectory, and is calculated as follows:

$$MSD(n\,\delta t) = \frac{1}{N-1-n} \sum_{j=1}^{N-1-n} \left[r\big((j+n)\,\delta t\big) - r\big(j\,\delta t\big) \right]^2$$

where $r(t)$ is the position of the molecule at time t during the trajectory, $n\,\delta t$ is the time lag at which the MSD is calculated, and δt is the smallest time interval resolved. The MSD is calculated for each trajectory to estimate the population MSD as a function of time. Plotting the MSD for each trajectory against time allows fitting to a theoretical model such as the Brownian diffusion equation (Michalet 2010; Ernst and Kohler 2013) and can yield information on the overall diffusion coefficients of the molecules, as well as identify deviation from the model, e.g., presence of any sub-diffusive or super-diffusive behaviour (Ernst and Kohler 2013).

- **Confinement analysis:** Given that biomolecules can experience confinement and corralling, it can be desirable to extract parts of trajectories that fit this behaviour (Simson, Sheets et al. 1995). Several analysis methods have been described for such situations. Commonly, these approaches instead of analysing the whole trajectory will sub-sample the trajectory with a defined step length, such as subdivision into equal parts (Serge, Bertaux et al. 2008). The diffusion properties for each section of the trajectory are then analysed, for example, using MSD analysis (see above; Shen, Tauzin et al. 2017). Parts of the track that deviate from Brownian diffusion and exhibit low MSD can then be used to identify confinements.

While the processing steps described here provide a general overview of the main methods and techniques used in SMT data processing, it is important to note that the specifics may vary depending on the system being studied, the imaging conditions, and the goals of the experiment.

3.5 ADVANCES AND EXTENSIONS OF FCS AND SPT

3.5.1 Fluorescence Imaging Correlation Spectroscopies

One of the main weaknesses of FCS is that it relies on single point measurements, as opposed to FRAP and SMT that can sample larger areas. This limitation has been attenuated by the implementation of image correlation spectroscopy (ICS) methods. ICS has been used in a wide range of biological applications, including the study of protein–protein interactions, cell membrane dynamics, and the study of actin assembly in living cells (Ruan, Cheng et al. 2004; Comeau, Kolin et al. 2008; Pandzic, Rossy et al. 2015). In addition, ICS has been used to study the behaviour of various systems at the molecular level, including lipid bilayers (Bag, Sankaran et al. 2012). Briefly, ICS works by measuring the fluctuations in pixel intensity between images of a sample taken at different times (Kolin and Wiseman 2007), and thus, the acquired data are like SMT experiments, i.e., an image stack. While ICS allows observation of larger regions of interest, it is limited to slow dynamics due to the time taken to acquire one image. Serving as a basis, ICS methods have led to natural extensions of FCS for probing molecular diffusion, whereby large regions can be probed, but the access to fast diffusion kinetics is maintained (Hebert, Costantino et al. 2005; Brown, Dalal et al. 2008).

Two prominent methods in this bracket are raster image correlation spectroscopy (RICS) and spatiotemporal correlation spectroscopy (STICS; Hebert, Costantino et al. 2005; Brown, Dalal et al. 2008). RICS exploits the hidden time structure within the raster scanning method common to CLSM. This allows rapid sampling (pixel dwell times ~μs) of many positions within the sample, thus allowing for extraction of fast dynamics from spatial correlation between positions on a scanline, such as those from FCS (Brown, Dalal et al. 2008). Slower dynamics are recovered from correlation information in successive lines and frames. RICS is now a well-established method and has been applied to numerous biological systems to measure the diffusion of proteins and lipids (Brown, Dalal et al. 2008; Digman and Gratton 2009; Gielen, Smisdom et al. 2009). STICS similarly exploits the time structure of raster scan image stacks, but applies filtering of immobile components, i.e., molecules that do not diffuse, in the Fourier space to allow measurement of subpopulations of diffusers within the sample (Hebert, Costantino et al. 2005). STICS also addresses a limitation of both RICS and FCS, as it can detect diffusion flow within the sample, owing to changes in the position of the 2D autocorrelation function. Thus, STICS has been implemented for probing the flow of the actin cytoskeleton (Ashdown, Pandzic et al. 2015; Ashdown and Owen 2018).

3.5.2 Super-Resolution FCS

One of the limitations of FCS is that the size of the observation volume can be too large to distinguish diffusion behaviour that occurs at the nanoscale or within sub-diffraction structures, e.g., lipid domains. Given super-resolution microscopy methods, such as stimulated emission depletion (STED) microscopy can access information on molecules at that scale, the method was recently extended to allow correlation spectroscopy measurements (Sezgin, Schneider et al. 2019). Here, the small size of the convolved

STED beam is exploited to give a small observation volume, approximately 30–50 nm in diameter. This has been used to probe diffusion within structures that are below the diffraction limit, particularly the diffusion of molecules in and out of small membrane domains (Honigmann, Mueller et al. 2014; Sezgin, Schneider et al. 2019). Honigmann et al. observed that lipid diffusion in the plasma membrane of living cells exhibited a complex diffusion behaviour, with a significant fraction of lipids showing restricted diffusion in submicron domains.

3.5.3 SMT at High Molecular Concentrations

As discussed earlier, SMT is severely limited by the number of fluorescent labels that can be active within a single frame of an acquisition, and this is normally determined by the molecular concentration. However, this limitation has been somewhat alleviated by the development of probes (Patterson and Lippincott-Schwartz 2002; Wiedenmann, Ivanchenko et al. 2004; Grimm, English et al. 2016; Grimm, Muthusamy et al. 2017) and techniques that allow subsampling of single molecule diffusion, e.g., points accumulation in nanoscale topography (PAINT; Sharonov and Hochstrasser 2006; Giannone, Hosy et al. 2013).

In the case of photoactivation experiments, early work focused on the use of newly developed photoactivatable proteins, such as PA-GFP and PA-mCherry, for SMT, referred to generally as sptPALM (Patterson and Lippincott-Schwartz 2002; Manley, Gillette et al. 2010). These probes are particularly useful for probing single molecule dynamics in biological assemblies normally too dense for SMT, e.g., focal adhesions (Rossier, Octeau et al. 2012). Furthermore, the development of new photoactivatable dyes, which are compatible with live cells, offer great potential for SMT. Recently developed Janelia Fluors (JFs) allow their fluorescence to be turned on with ~400 nm excitation (Grimm, English et al. 2016), have already been demonstrated for SMT, and are available currently available in two distinct wavelengths (i.e., PA-JF549 and PA-JF646). Thus, these probes could allow SMT of two proteins at high concentration simultaneously in the future, a problem currently beyond most SMT methods.

The development of PAINT-based approaches has given another route for achieving SMT in high molecule densities (Sharonov and Hochstrasser 2006; Giannone, Hosy et al. 2010; Jungmann, Steinhauer et al. 2010). Briefly, PAINT approaches work by having an excess of fluorescent label in solution, with the fluorophore normally attached to an antibody or small binder (e.g., receptor ligand or DNA strand). The unbound fluorophores diffuse too quickly in solution to give distinct PSFs within the data, but it is worthy to note that the overall background in the images is increased (Giannone, Hosy et al. 2010). Upon binding of the fluorophore labelled binder to its target, it is now slow enough to be detected above background under oblique illumination. This means that for the time the label is bound, the molecule can be imaged and tracked, and if the labels can dissociate, this process can be repeated for long observation times by exploiting the pool of excess label (Giannone, Hosy et al. 2010). Furthermore, the average density of molecules "on" can be tuned by the concentration of the unbound fluorophore used, with dependence on the binding kinetics of the label to the target. This approach is limited to extracellular targets, given the requirement for external labels, but it has been demonstrated for SMT of several cell surface receptors (e.g., AMPA, EGFR; Giannone, Hosy et al. 2010; Winckler, Lartigue et al. 2013).

3.6 INSIGHTS INTO BIOMOLECULAR REGULATION AND BEHAVIOUR USING FCS AND SPT

3.6.1 Neuroscience – AMPA Receptor Membrane Dynamics in Neurons

AMPA receptors are ionotropic glutamate receptors that mediate fast excitatory synaptic transmission in the central nervous system. They play an important role in neuronal functions and are responsible for most synaptic transmission throughout the nervous system (Henley and Wilkinson 2016). Control and regulation of AMPA receptor behaviour underlies the large flexibility in neuronal activity in the central nervous system. Dysregulation of AMPA receptor function can lead to excessive glutamate release, resulting in neuronal damage or death, and is implicated in various neurological diseases, including epilepsy, stroke, and Alzheimer's disease (Henley and Wilkinson 2016). Given that AMPA receptors are a promising target for pharmaceutical intervention in these diseases, there is great interest in determining how they are regulated within the plasma membrane (Anggono and Huganir 2012). There has therefore been great focus on single molecule observations of AMPA receptors; in particular, SMT measurements in neurons have given new insights into regulation of single AMPA receptors (Tardin, Cognet et al. 2003; Hoze, Nair et al. 2012; Nair, Hosy et al. 2013).

Monomeric AMPA receptors were initially observed by SMT studies to be highly mobile and can rapidly diffuse within the synaptic membrane (diffusion coefficients ranging from 0.04 to 0.2 μm^2/s; Tardin, Cognet et al. 2003; Morise, Suzuki et al. 2019). Furthermore, their turnover is highly dependent on the diffusion rate of individual receptors, suggesting that monomer diffusion plays a key role in the regulation of synaptic strength (Morise, Suzuki et al. 2019). The effect of glutamate stimulation on AMPA receptor lateral diffusion showed that binding of glutamate can regulate their lateral movement (Morise, Suzuki et al. 2019). While glutamate stimulation has been observed to give a transient increase in the mobility of AMPA receptors, with diffusion coefficients increasing by 15%–30%, depending on activation of NMDA receptors and the influx of calcium ions, it has also been shown that eventually the overall diffusion of the receptors is decreased due to the activity of protein kinase C (Tardin, Cognet et al. 2003; Morise, Suzuki et al. 2019).

Overall, SMT methods have provided new observations on the lateral diffusion, regulation, and trafficking of AMPA receptors at the membrane, and this is regulated by a complex interplay of molecular mechanisms, including the presence of other proteins, the activity of the actin cytoskeleton, and the activity of specific enzymes. These findings will have important implications for our understanding of the regulation of synaptic transmission and plasticity, as well as for the development of new treatments for neurological disorders that involve dysfunction of AMPA receptor signalling.

3.6.2 Immunology – T-Cell Receptor Engagement and Triggering

One of the critical questions in immunology is how small amounts of antigenic material, present on the surface of antigen-presenting cells, is detected and initiates potent responses in T-cells, and furthermore, how T-cells discriminate between these antigens. Several models have been proposed for this process (Hopfield 1974; Valitutti, Muller et al. 1995; Davis and van der Merwe 2006; Huang, Brameshuber et al. 2013), and due to the use of single molecule methods, great insights have been gained into how these models proceed in the context of the immune synapse (Dustin and Depoil 2011; O'Donoghue, Pielak et al. 2013; Pageon, Tabarin et al. 2016).

It has been proposed that for T-cells to discriminate between different antigens presented by antigen-presenting cells, the diffusion and spatial organisation of TCRs and a phosphatase, CD45, is critical (Furlan, Minowa et al. 2014). Initially, work focused on the movement of TCR "microclusters" at the immune synapse (DeMond, Mossman et al. 2008), but has since evolved into probing the movement of single TCRs (O'Donoghue,

Pielak et al. 2013; Ponjavic, McColl et al. 2018; Nieves, Pandzic et al. 2022). SMT was used to probe TCR triggering using fluorescently labelled antigen and TCR. It was found that the diffusion of engaged TCR was consistent with the binding of single antigens and was dependent on the strength of the TCR-peptide-major histocompatibility complex (pMHC) interaction (O'Donoghue, Pielak et al. 2013). Sustained triggering led to the formation of TCR microclusters, which were dependent on the strength of the TCR-pMHC interaction and the presence of co-stimulatory molecules. Furthermore, the kinetics of TCR triggering were influenced by the density of pMHC on the APC surface, with higher densities leading to more rapid and sustained TCR triggering, via increased immobilisation of the receptor (O'Donoghue, Pielak et al. 2013).

3.6.3 Cancer – Epidermal Growth Factor Receptor Assembly Dynamics and Signalling Amplification

Epidermal growth factor receptor (EGFR) is a transmembrane receptor that plays a crucial role in the regulation of cell proliferation, differentiation, and survival. Dysregulation of EGFR signalling has been implicated in the development and progression of a wide range of human cancers, including lung, breast, colorectal, and head and neck cancers (Han and Lo 2012). Thus, EGFR signalling is a target of great interest for therapeutics in cancer treatment, with several EGFR inhibitors already developed and approved for clinical use (Han and Lo 2012).

Single molecule studies have been employed extensively to observe the diffusion of monomeric EGFR within the membrane and response to ligand binding (Chung, Akita et al. 2010; Ibach, Radon et al. 2015; Kim, Kim et al. 2017). FCS showed that upon activation by epidermal growth factor (EGF), the EGFR undergoes a significant reduction in mobility. Specifically, the diffusion coefficient decreased from 0.33 to 0.07 $\mu m^2/s$ (Kim, Kim et al. 2017), indicating that the receptor becomes more immobilised upon activation, as observed in several SMT studies (Chung, Akita et al. 2010; Ibach, Radon et al. 2015; Kim, Kim et al. 2017). This would be consistent with the formation of EGFR dimers and oligomers at the membrane. This was investigated by Liu et al. using fluorescence cross-correlation spectroscopy, where the fraction of dimers of EGFR and HER2, another HER family member, was measured. It was observed that receptors pre-formed into dimers, but EGFR-HER2 heterodimerisation occurred more frequently than homodimers. Dimerisation behaviour was also found to depend on ligand binding and the identity of proteins in dimers; for example, dimerisation for HER2 was found to be independent of ligand binding, while dimerisation for HER3 required ligand binding (Liu, Sudhaharan et al. 2007).

EGFR dimers exhibited distinct spatial distributions depending on their activation status (Chung, Akita et al. 2010), and rapid internalisation is a mechanism for amplified signalling (Ibach, Radon et al. 2015). Activated EGFR dimers were found to be highly clustered in regions of the plasma membrane that were rich in extracellular matrix proteins, whereas inactive EGFR dimers were more uniformly distributed across the plasma membrane (Chung, Akita et al. 2010). Furthermore, reversible EGFR dimerisation played a key role in the spatial control of EGFR activation. Specifically, the dimerisation of inactive EGFR dimers was reversible, and these dimers were capable of being recruited into active signalling complexes in response to EGF stimulation (Chung, Akita et al. 2010).

Overall, the single molecule methods have provided valuable insights into the spatial control of EGFR activation by reversible dimerisation and internalisation on living cells. The spatial distribution and diffusion of EGFR monomers and subsequent dimers is tightly regulated by within the plasma membrane.

REFERENCES

Ando, R., H. Hama, M. Yamamoto-Hino, H. Mizuno and A. Miyawaki (2002). "An optical marker based on the UV-induced green-to-red photoconversion of a fluorescent protein." *Proceedings of the National Academy of Sciences of the United States of America* **99**(20): 12651–12656.

Anggono, V. and R. L. Huganir (2012). "Regulation of AMPA receptor trafficking and synaptic plasticity." *Current Opinion in Neurobiology* **22**(3): 461–469.

Ashdown, G., E. Pandzic, A. Cope, P. Wiseman and D. Owen (2015). "Cortical actin flow in T cells quantified by spatio-temporal image correlation spectroscopy of structured illumination microscopy data." *Journal of Visualized Experiments* **2015**(106): e53749.

Ashdown, G. W. and D. M. Owen (2018). "Spatio-temporal image correlation spectroscopy and super-resolution microscopy to quantify molecular dynamics in T cells." *Methods* **140–141**: 112–118.

Axelrod, D., D. E. Koppel, J. Schlessinger, E. Elson and W. W. Webb (1976). "Mobility measurement by analysis of fluorescence photobleaching recovery kinetics." *Biophysical Journal* **16**(9): 1055–1069.

Bachmann, M., S. Kukkurainen, V. P. Hytonen and B. Wehrle-Haller (2019). "Cell adhesion by integrins." *Physiological Reviews* **99**(4): 1655–1699.

Bacia, K., D. Scherfeld, N. Kahya and P. Schwille (2004). "Fluorescence correlation spectroscopy relates rafts in model and native membranes." *Biophysical Journal* **87**(2): 1034–1043.

Bag, N., J. Sankaran, A. Paul, R. S. Kraut and T. Wohland (2012). "Calibration and limits of camera-based fluorescence correlation spectroscopy: a supported lipid bilayer study." *Chemphyschem* **13**(11): 2784–2794.

Balzarotti, F., Y. Eilers, K. C. Gwosch, A. H. Gynna, V. Westphal, F. D. Stefani, J. Elf and S. W. Hell (2017). "Nanometer resolution imaging and tracking of fluorescent molecules with minimal photon fluxes." *Science* **355**(6325): 606–612.

Beckers, D., D. Urbancic and E. Sezgin (2020). "Impact of nanoscale hindrances on the relationship between lipid packing and diffusion in model membranes." *The Journal of Physical Chemistry B* **124**(8): 1487–1494.

Benke, A., N. Olivier, J. Gunzenhauser and S. Manley (2012). "Multicolor single molecule tracking of stochastically active synthetic dyes." *Nano Letters* **12**(5): 2619–2624.

Berland, K. M., P. T. So and E. Gratton (1995). "Two-photon fluorescence correlation spectroscopy: method and application to the intracellular environment." *Biophysical Journal* **68**(2): 694–701.

Betzig, E., G. H. Patterson, R. Sougrat, O. W. Lindwasser, S. Olenych, J. S. Bonifacino, M. W. Davidson, J. Lippincott-Schwartz and H. F. Hess (2006). "Imaging intracellular fluorescent proteins at nanometer resolution." *Science* **313**(5793): 1642–1645.

Blom, H., L. Kastrup and C. Eggeling (2006). "Fluorescence fluctuation spectroscopy in reduced detection volumes." *Current Pharmaceutical Biotechnology* **7**(1): 51–66.

Brown, C. M., R. B. Dalal, B. Hebert, M. A. Digman, A. R. Horwitz and E. Gratton (2008). "Raster image correlation spectroscopy (RICS) for measuring fast protein dynamics and concentrations with a commercial laser scanning confocal microscope." *Journal of Microscopy* **229**(Pt 1): 78–91.

Brown, G. C. and B. N. Kholodenko (1999). "Spatial gradients of cellular phospho-proteins." *FEBS Letters* **457**(3): 452–454.

Buchholz, J., J. Krieger, C. Bruschini, S. Burri, A. Ardelean, E. Charbon and J. Langowski (2018). "Widefield high frame rate single-photon SPAD imagers for SPIM-FCS." *Biophysical Journal* **114**(10): 2455–2464.

Buckley, C. D., G. E. Rainger, P. F. Bradfield, G. B. Nash and D. L. Simmons (1998). "Cell adhesion: more than just glue (review)." *Molecular Membrane Biology* **15**(4): 167–176.

Cavalcanti-Adam, E. A., A. Micoulet, J. Blummel, J. Auernheimer, H. Kessler and J. P. Spatz (2006). "Lateral spacing of integrin ligands influences cell spreading and focal adhesion assembly." *European Journal of Cell Biology* **85**(3–4): 219–224.

Cavalcanti-Adam, E. A., T. Volberg, A. Micoulet, H. Kessler, B. Geiger and J. P. Spatz (2007). "Cell spreading and focal adhesion dynamics are regulated by spacing of integrin ligands." *Biophysical Journal* **92**(8): 2964–2974.

Cheezum, M. K., W. F. Walker and W. H. Guilford (2001). "Quantitative comparison of algorithms for tracking single fluorescent particles." *Biophysical Journal* **81**(4): 2378–2388.

Chung, I., R. Akita, R. Vandlen, D. Toomre, J. Schlessinger and I. Mellman (2010). "Spatial control of EGF receptor activation by reversible dimerization on living cells." *Nature* **464**(7289): 783–787.

Comeau, J. W., D. L. Kolin and P. W. Wiseman (2008). "Accurate measurements of protein interactions in cells via improved spatial image cross-correlation spectroscopy." *Molecular BioSystems* **4**(6): 672–685.

Cordoba, S. P., K. Choudhuri, H. Zhang, M. Bridge, A. B. Basat, M. L. Dustin and P. A. van der Merwe (2013). "The large ectodomains of CD45 and CD148 regulate their segregation from and inhibition of ligated T-cell receptor." *Blood* **121**(21): 4295–4302.

Davis, S. J. and P. A. van der Merwe (2006). "The kinetic-segregation model: TCR triggering and beyond." *Nature Immunology* **7**(8): 803–809.

DeMond, A. L., K. D. Mossman, T. Starr, M. L. Dustin and J. T. Groves (2008). "T cell receptor microcluster transport through molecular mazes reveals mechanism of translocation." *Biophysical Journal* **94**(8): 3286–3292.

Digman, M. A. and E. Gratton (2009). "Analysis of diffusion and binding in cells using the RICS approach." *Microscopy Research and Technique* **72**(4): 323–332.

Duchesne, L., V. Octeau, R. N. Bearon, A. Beckett, I. A. Prior, B. Lounis and D. G. Fernig (2012). "Transport of fibroblast growth factor 2 in the pericellular matrix is controlled by the spatial distribution of its binding sites in heparan sulfate." *PLoS Biology* **10**(7): e1001361.

Dustin, M. L. and D. Depoil (2011). "New insights into the T cell synapse from single molecule techniques." *Nature Reviews Immunology* **11**(10): 672–684.

Elliott, A. D. (2020). "Confocal microscopy: principles and modern practices." *Current Protocols in Cytometry* **92**(1): e68.

Ernst, D. and J. Kohler (2013). "Measuring a diffusion coefficient by single-particle tracking: statistical analysis of experimental mean squared displacement curves." *Physical Chemistry Chemical Physics* **15**(3): 845–849.

Felsenfeld, D. P., D. Choquet and M. P. Sheetz (1996). "Ligand binding regulates the directed movement of beta1 integrins on fibroblasts." *Nature* **383**(6599): 438–440.

Fernandez-Suarez, M. and A. Y. Ting (2008). "Fluorescent probes for super-resolution imaging in living cells." *Nature Reviews Molecular Cell Biology* **9**(12): 929–943.

Furlan, G., T. Minowa, N. Hanagata, C. Kataoka-Hamai and Y. Kaizuka (2014). "Phosphatase CD45 both positively and negatively regulates T cell receptor phosphorylation in reconstituted membrane protein clusters." *Journal of Biological Chemistry* **289**(41): 28514–28525.

Giannone, G., E. Hosy, F. Levet, A. Constals, K. Schulze, A. I. Sobolevsky, M. P. Rosconi, E. Gouaux, R. Tampe, D. Choquet and L. Cognet (2010). "Dynamic superresolution imaging of endogenous proteins on living cells at ultra-high density." *Biophysical Journal* **99**(4): 1303–1310.

Giannone, G., E. Hosy, J. B. Sibarita, D. Choquet and L. Cognet (2013). "High-content super-resolution imaging of live cell by uPAINT." *Methods in Molecular Biology* **950**: 95–110.

Gielen, E., N. Smisdom, M. vandeVen, B. De Clercq, E. Gratton, M. Digman, J. M. Rigo, J. Hofkens, Y. Engelborghs and M. Ameloot (2009). "Measuring diffusion of lipid-like probes in artificial and natural membranes by raster image correlation spectroscopy (RICS): use of a commercial laser-scanning microscope with analog detection." *Langmuir* **25**(9): 5209–5218.

Goehring, N. W., D. Chowdhury, A. A. Hyman and S. W. Grill (2010). "FRAP analysis of membrane-associated proteins: lateral diffusion and membrane-cytoplasmic exchange." *Biophysical Journal* **99**(8): 2443–2452.

Grimm, J. B., B. P. English, H. Choi, A. K. Muthusamy, B. P. Mehl, P. Dong, T. A. Brown, J. Lippincott-Schwartz, Z. Liu, T. Lionnet and L. D. Lavis (2016). "Bright photoactivatable fluorophores for single-molecule imaging." *Nature Methods* **13**(12): 985–988.

Grimm, J. B., A. K. Muthusamy, Y. Liang, T. A. Brown, W. C. Lemon, R. Patel, R. Lu, J. J. Macklin, P. J. Keller, N. Ji and L. D. Lavis (2017). "A general method to fine-tune fluorophores for live-cell and in vivo imaging." *Nature Methods* **14**(10): 987–994.

Gurevich, V. V. and E. V. Gurevich (2019). "GPCR signaling regulation: the role of GRKs and arrestins." *Frontiers in Pharmacology* **10**: 125.

Han, W. and H. W. Lo (2012). "Landscape of EGFR signaling network in human cancers: biology and therapeutic response in relation to receptor subcellular locations." *Cancer Letters* **318**(2): 124–134.

Harayama, T. and H. Riezman (2018). "Understanding the diversity of membrane lipid composition." *Nature Reviews Molecular Cell Biology* **19**(5): 281–296.

Haustein, E. and P. Schwille (2007). "Trends in fluorescence imaging and related techniques to unravel biological information." *HFSP Journal* **1**(3): 169–180.

Hebert, B., S. Costantino and P. W. Wiseman (2005). "Spatiotemporal image correlation spectroscopy (STICS) theory, verification, and application to protein velocity mapping in living CHO cells." *Biophysical Journal* **88**(5): 3601–3614.

Heim, R., D. C. Prasher and R. Y. Tsien (1994). "Wavelength mutations and posttranslational autoxidation of green fluorescent protein." *Proceedings of the National Academy of Sciences of the United States of America* **91**(26): 12501–12504.

Hell, S. W. (2009). "Microscopy and its focal switch." *Nature Methods* **6**(1): 24–32.

Henley, J. M. and K. A. Wilkinson (2016). "Synaptic AMPA receptor composition in development, plasticity and disease." *Nature Reviews Neuroscience* **17**(6): 337–350.

Heuff, R. F., J. L. Swift and D. T. Cramb (2007). "Fluorescence correlation spectroscopy using quantum dots: advances, challenges and opportunities." *Physical Chemistry Chemical Physics* **9**(16): 1870–1880.

Hilzenrat, G., E. Pandzic, Z. Yang, D. J. Nieves, J. Goyette, J. Rossy, Y. Ma and K. Gaus (2020). "Conformational states control lck switching between free and confined diffusion modes in T cells." *Biophysical Journal* **118**(6): 1489–1501.

Hodges, C., R. P. Kafle, J. D. Hoff and J. C. Meiners (2018). "Fluorescence correlation spectroscopy with photobleaching correction in slowly diffusing systems." *Journal of Fluorescence* **28**(2): 505–511.

Honigmann, A., V. Mueller, H. Ta, A. Schoenle, E. Sezgin, S. W. Hell and C. Eggeling (2014). "Scanning STED-FCS reveals spatiotemporal heterogeneity of lipid interaction in the plasma membrane of living cells." *Nature Communications* **5**: 5412.

Hopfield, J. J. (1974). "Kinetic proofreading: a new mechanism for reducing errors in biosynthetic processes requiring high specificity." *Proceedings of the National Academy of Sciences of the United States of America* **71**(10): 4135–4139.

Hoze, N., D. Nair, E. Hosy, C. Sieben, S. Manley, A. Herrmann, J. B. Sibarita, D. Choquet and D. Holcman (2012). "Heterogeneity of AMPA receptor trafficking and molecular interactions revealed by superresolution analysis of live cell imaging." *Proceedings of the National Academy of Sciences of the United States of America* **109**(42): 17052–17057.

Huang, J., M. Brameshuber, X. Zeng, J. Xie, Q. J. Li, Y. H. Chien, S. Valitutti and M. M. Davis (2013). "A single peptide-major histocompatibility complex ligand triggers digital cytokine secretion in CD4(+) T cells." *Immunity* **39**(5): 846–857.

Ibach, J., Y. Radon, M. Gelléri, M. H. Sonntag, L. Brunsveld, P. I. H. Bastiaens and P. J. Verveer (2015). "Single particle tracking reveals that EGFR signaling activity is amplified in clathrin-coated pits." *PLOS One* **10**(11): e0143162.

Irles, C., A. Symons, F. Michel, T. R. Bakker, P. A. van der Merwe and O. Acuto (2003). "CD45 ectodomain controls interaction with GEMs and Lck activity for optimal TCR signaling." *Nature Immunology* **4**: 189–197.

Iyer, G., X. Michalet, Y. P. Chang and S. Weiss (2010). "Tracking single proteins in live cells using single-chain antibody fragment-fluorescent quantum dot affinity pair." *Methods in Enzymology* **475**: 61–79.

Juillerat, A., T. Gronemeyer, A. Keppler, S. Gendreizig, H. Pick, H. Vogel and K. Johnsson (2003). "Directed evolution of O6-alkylguanine-DNA alkyltransferase for efficient labeling of fusion proteins with small molecules in vivo." *Chemical Biology* **10**(4): 313–317.

Jungmann, R., C. Steinhauer, M. Scheible, A. Kuzyk, P. Tinnefeld and F. C. Simmel (2010). "Single-molecule kinetics and super-resolution microscopy by fluorescence imaging of transient binding on DNA origami." *Nano Letters* **10**(11): 4756–4761.

Kholodenko, B. N. (2006). "Cell-signalling dynamics in time and space." *Nature Reviews Molecular Cell Biology* **7**(3): 165–176.

Kim, D. H., D. K. Kim, K. Zhou, S. Park, Y. Kwon, M. G. Jeong, N. K. Lee and S. H. Ryu (2017). "Single particle tracking-based reaction progress kinetic analysis reveals a series of molecular mechanisms of cetuximab-induced EGFR processes in a single living cell." *Chemical Science* **8**(7): 4823–4832.

Kim, S. A., K. G. Heinze and P. Schwille (2007). "Fluorescence correlation spectroscopy in living cells." *Nature Methods* **4**(11): 963–973.

Kinoshita, M., K. G. N. Suzuki, M. Murata and N. Matsumori (2018). "Evidence of lipid rafts based on the partition and dynamic behavior of sphingomyelins." *Chemistry and Physics of Lipids* **215**: 84–95.

Kolin, D. L. and P. W. Wiseman (2007). "Advances in image correlation spectroscopy: measuring number densities, aggregation states, and dynamics of fluorescently labeled macromolecules in cells." *Cell Biochemistry and Biophysics* **49**(3): 141–164.

Kusumi, A., Y. Sako and M. Yamamoto (1993). "Confined lateral diffusion of membrane receptors as studied by single particle tracking (nanovid microscopy). Effects of calcium-induced differentiation in cultured epithelial cells." *Biophysical Journal* **65**(5): 2021–2040.

Lippincott-Schwartz, J., E. L. Snapp and R. D. Phair (2018). "The development and enhancement of FRAP as a key tool for investigating protein dynamics." *Biophysical Journal* **115**(7): 1146–1155.

Liu, P., T. Sudhaharan, R. M. Koh, L. C. Hwang, S. Ahmed, I. N. Maruyama and T. Wohland (2007). "Investigation of the dimerization of proteins from the epidermal growth factor receptor family by single wavelength fluorescence cross-correlation spectroscopy." *Biophysical Journal* **93**(2): 684–698.

Los, G. V., L. P. Encell, M. G. McDougall, D. D. Hartzell, N. Karassina, C. Zimprich, M. G. Wood, R. Learish, R. F. Ohana, M. Urh, D. Simpson, J. Mendez, K. Zimmerman, P. Otto, G. Vidugiris, J. Zhu, A. Darzins, D. H. Klaubert, R. F. Bulleit and K. V. Wood (2008). "HaloTag: a novel protein labeling technology for cell imaging and protein analysis." ACS *Chemical Biology* **3**(6): 373–382.

Magde, D., E. Elson and W. W. Webb (1972). "Thermodynamic fluctuations in a reacting system---measurement by fluorescence correlation spectroscopy." *Physical Review Letters* **29**(11): 705–708.

Manley, S., J. M. Gillette and J. Lippincott-Schwartz (2010). "Single-particle tracking photoactivated localization microscopy for mapping single-molecule dynamics." *Methods Enzymol* **475**: 109–120.

Manzo, C. and M. F. Garcia-Parajo (2015). "A review of progress in single particle tracking: from methods to biophysical insights." *Reports on Progress in Physics* **78**(12): 124601.

Martens, K. J. A., B. Turkowyd and U. Endesfelder (2021). "Raw data to results: a hands-on introduction and overview of computational analysis for single-molecule localization microscopy." *Frontiers in Bioinformatics* **1**: 817254.

Michalet, X. (2010). "Mean square displacement analysis of single-particle trajectories with localization error: Brownian motion in an isotropic medium." *Physical Review E, Statistical, Nonlinear, and Soft Matter Physics* **82**(4 Pt 1): 041914.

Morise, J., K. G. N. Suzuki, A. Kitagawa, Y. Wakazono, K. Takamiya, T. A. Tsunoyama, Y. L. Nemoto, H. Takematsu, A. Kusumi and S. Oka (2019). "AMPA receptors in the synapse turnover by monomer diffusion." *Nature Communications* **10**(1): 5245.

Mudumbi, K. C., E. C. Schirmer and W. Yang (2016). "Single-point single-molecule FRAP distinguishes inner and outer nuclear membrane protein distribution." *Nature Communications* **7**: 12562.

Nair, D., E. Hosy, J. D. Petersen, A. Constals, G. Giannone, D. Choquet and J. B. Sibarita (2013). "Super-resolution imaging reveals that AMPA receptors inside synapses are dynamically organized in nanodomains regulated by PSD95." *The Journal of Neuroscience* **33**(32): 13204–13224.

Nicolson, G. L. and G. Ferreira de Mattos (2022). "Fifty years of the fluid-mosaic model of biomembrane structure and organization and its importance in biomedicine with particular emphasis on membrane lipid replacement." *Biomedicines* **10**(7): 1–25.

Nieves, D. J. and M. A. B. Baker (2021). "Pushing the super-resolution limit: recent improvements in microscopy below the diffraction limit." *Biochemical Society Transactions* **49**(1): 431–439.

Nieves, D. J., E. Pandzic, S. D. Gunasinghe, J. Goyette, D. M. Owen, J. Justin Gooding and K. Gaus (2022). "The T cell receptor displays lateral signal propagation involving non-engaged receptors." *Nanoscale* **14**(9): 3513–3526.

Nikic, I., J. H. Kang, G. E. Girona, I. V. Aramburu and E. A. Lemke (2015). "Labeling proteins on live mammalian cells using click chemistry." *Nature Protocols* **10**(5): 780–791.

O'Donoghue, G. P., R. M. Pielak, A. A. Smoligovets, J. J. Lin and J. T. Groves (2013). "Direct single molecule measurement of TCR triggering by agonist pMHC in living primary T cells." *Elife* **2**: e00778.

Oshima, A., H. Nakashima and K. Sumitomo (2019). "Evaluation of lateral diffusion of lipids in continuous membranes between freestanding and supported areas by fluorescence recovery after photobleaching." *Langmuir* **35**(36): 11725–11734.

Ostman, A. and F. D. Bohmer (2001). "Regulation of receptor tyrosine kinase signaling by protein tyrosine phosphatases." *Trends in Cell Biology* **11**(6): 258–266.

Pageon, S. V., T. Tabarin, Y. Yamamoto, Y. Ma, J. S. Bridgeman, A. Cohnen, C. Benzing, Y. Gao, M. D. Crowther, K. Tungatt, G. Dolton, A. K. Sewell, D. A. Price, O. Acuto, R. G. Parton, J. J. Gooding, J. Rossy, J. Rossjohn and K. Gaus (2016). "Functional role of T-cell receptor nanoclusters in signal initiation and antigen discrimination." *Proceedings of the National Academy of Sciences of the United States of America* **113**(37): E5454–5463.

Pandzic, E., J. Rossy and K. Gaus (2015). "Tracking molecular dynamics without tracking: image correlation of photo-activation microscopy." *Methods and Applications in Fluorescence* **3**(1): 014006.

Patterson, G. H. and J. Lippincott-Schwartz (2002). "A photoactivatable GFP for selective photolabeling of proteins and cells." *Science* **297**(5588): 1873–1877.

Pettmann, J., A. Huhn, E. Abu Shah, M. A. Kutuzov, D. B. Wilson, M. L. Dustin, S. J. Davis, P. A. van der Merwe and O. Dushek (2021). "The discriminatory power of the T cell receptor." *Elife* **10**: 1–42.

Ponjavic, A., J. McColl, A. R. Carr, A. M. Santos, K. Kulenkampff, A. Lippert, S. J. Davis, D. Klenerman and S. F. Lee (2018). "Single-molecule light-sheet imaging of suspended t cells." *Biophysical Journal* **114**(9): 2200–2211.

Razvag, Y., Y. Neve-Oz, J. Sajman, M. Reches and E. Sherman (2018). "Nanoscale kinetic segregation of TCR and CD45 in engaged microvilli facilitates early T cell activation." *Nature Communications* **9**(1): 732.

Ritchie, K., X. Y. Shan, J. Kondo, K. Iwasawa, T. Fujiwara and A. Kusumi (2005). "Detection of non-Brownian diffusion in the cell membrane in single molecule tracking." *Biophysical Journal* **88**(3): 2266–2277.

Rossier, O., V. Octeau, J. B. Sibarita, C. Leduc, B. Tessier, D. Nair, V. Gatterdam, O. Destaing, C. Albiges-Rizo, R. Tampe, L. Cognet, D. Choquet, B. Lounis and G. Giannone (2012). "Integrins beta1 and beta3 exhibit distinct dynamic nanoscale organizations inside focal adhesions." *Nature Cell Biology* **14**(10): 1057–1067.

Ruan, Q., M. A. Cheng, M. Levi, E. Gratton and W. W. Mantulin (2004). "Spatial-temporal studies of membrane dynamics: scanning fluorescence correlation spectroscopy (SFCS)." *Biophysical Journal* **87**(2): 1260–1267.

Ruttinger, S., V. Buschmann, B. Kramer, R. Erdmann, R. Macdonald and F. Koberling (2008). "Comparison and accuracy of methods to determine the confocal volume for quantitative fluorescence correlation spectroscopy." *Journal of Microscopy* **232**(2): 343–352.

Sako, Y. and A. Kusumi (1994). "Compartmentalized structure of the plasma membrane for receptor movements as revealed by a nanometer-level motion analysis." *Journal of Cell Biology* **125**(6): 1251–1264.

Saxton, M. J. (2008). "Single-particle tracking: connecting the dots." *Nature Methods* **5**(8): 671–672.

Saxton, M. J. and K. Jacobson (1997). "Single-particle tracking: applications to membrane dynamics." *Annual Review of Biophysics and Biomolecular Structure* **26**: 373–399.

Schlessinger, J. (2000). "Cell signaling by receptor tyrosine kinases." *Cell* **103**(2): 211–225.

Schwille, P. (2001). "Fluorescence correlation spectroscopy and its potential for intracellular applications." *Cell Biochemistry and Biophysics* **34**(3): 383–408.

Schwille, P., U. Haupts, S. Maiti and W. W. Webb (1999). "Molecular dynamics in living cells observed by fluorescence correlation spectroscopy with one- and two-photon excitation." *Biophysical Journal* **77**(4): 2251–2265.

Serge, A., N. Bertaux, H. Rigneault and D. Marguet (2008). "Dynamic multiple-target tracing to probe spatiotemporal cartography of cell membranes." *Nature Methods* **5**(8): 687–694.

Sezgin, E., F. Schneider, S. Galiani, I. Urbancic, D. Waithe, B. C. Lagerholm and C. Eggeling (2019). "Measuring nanoscale diffusion dynamics in cellular membranes with super-resolution STED-FCS." *Nature Protocols* **14**(4): 1054–1083.

Shannon, M. J., J. Pineau, J. Griffie, J. Aaron, T. Peel, D. J. Williamson, R. Zamoyska, A. P. Cope, G. H. Cornish and D. M. Owen (2019). "Differential nanoscale organisation of LFA-1 modulates T-cell migration." *Journal of Cell Science* **133**(5): 1–12.

Sharonov, A. and R. M. Hochstrasser (2006). "Wide-field subdiffraction imaging by accumulated binding of diffusing probes." *Proceedings of the National Academy of Sciences of the United States of America* **103**(50): 18911–18916.

Shen, H., L. J. Tauzin, R. Baiyasi, W. Wang, N. Moringo, B. Shuang and C. F. Landes (2017). "Single particle tracking: from theory to biophysical applications." *Chemical Reviews* **117**(11): 7331–7376.

Simson, R., E. D. Sheets and K. Jacobson (1995). "Detection of temporary lateral confinement of membrane proteins using single-particle tracking analysis." *Biophysical Journal* **69**(3): 989–993.

Sun, C., M. Marcello, Y. Li, D. Mason, R. Levy and D. G. Fernig (2016). "Selectivity in glycosaminoglycan binding dictates the distribution and diffusion of fibroblast growth factors in the pericellular matrix." *Open Biology* **6**(3). https://doi.org/10.1098/rsob.150277

Tardin, C., L. Cognet, C. Bats, B. Lounis and D. Choquet (2003). "Direct imaging of lateral movements of AMPA receptors inside synapses." *EMBO Journal* **22**(18): 4656–4665.

Tsien, R. Y. (1998). "The green fluorescent protein." *Annual Review of Biochemistry* **67**: 509–544.

Valitutti, S., S. Muller, M. Cella, E. Padovan and A. Lanzavecchia (1995). "Serial triggering of many T-cell receptors by a few peptide-MHC complexes." *Nature* **375**(6527): 148–151.

Walder, R., M. Kastantin and D. K. Schwartz (2012). "High throughput single molecule tracking for analysis of rare populations and events." *Analyst* **137**(13): 2987–2996.

Wiedenmann, J., S. Ivanchenko, F. Oswald, F. Schmitt, C. Rocker, A. Salih, K. D. Spindler and G. U. Nienhaus (2004). "EosFP, a fluorescent marker protein with UV-inducible green-to-red fluorescence conversion." *Proceedings of the National Academy of Sciences of the United States of America* **101**(45): 15905–15910.

Wieser, S. and G. J. Schutz (2008). "Tracking single molecules in the live cell plasma membrane-Do's and Don't's." *Methods* **46**(2): 131–140.

Winckler, P., L. Lartigue, G. Giannone, F. De Giorgi, F. Ichas, J. B. Sibarita, B. Lounis and L. Cognet (2013). "Identification and super-resolution imaging of ligand-activated receptor dimers in live cells." *Scientific Reports* **3**: 2387.

Yauch, R. L., D. P. Felsenfeld, S. K. Kraeft, L. B. Chen, M. P. Sheetz and M. E. Hemler (1997). "Mutational evidence for control of cell adhesion through integrin diffusion/clustering, independent of ligand binding." *Journal of Experimental Medicine* **186**(8): 1347–1355.

Yu, L., Y. Lei, Y. Ma, M. Liu, J. Zheng, D. Dan and P. Gao (2021). "A comprehensive review of fluorescence correlation spectroscopy." *Frontiers in Physics* **9**: 644450.

CHAPTER 4

Detecting Membrane Protein Dimers and Oligomers by FRET

4.1 INTRODUCTION TO MEMBRANE DIMERISATION

The dimerisation of membrane proteins can occur through a variety of mechanisms (Owen, Williamson et al. 2009), including coiled-coil interactions, transmembrane domain interactions, and intracellular/extracellular domain interactions. For receptors, ligand-induced dimerisation is also possible. Coiled-coil interactions involve the formation of an alphahelical structure in which two identical or similar proteins wrap around each other. Transmembrane domain interactions involve the association of two transmembrane domains, either through hydrophobic interactions or through the formation of a beta-sheet structure. Extracellular or intracellular domain interactions involve the association of two intracellular/extracellular domains, often through the formation of disulphide bonds, for example, in the case of the T cell receptor zeta-chains (Orloff, Ra et al. 1990).

The dimerisation of membrane proteins can have a significant impact on their structure and function. For example, the dimerisation of some ion channels, such as the voltage-gated sodium channel, is necessary for their proper function (Clatot, Hoshi et al. 2017). The T cell receptor is another classic example of dimerisation, with the majority of T cells expressing the α/β heterodimer which forms a functional receptor unit (Wang, Lim et al.

DOI: 10.1201/9781003273745-4

1998), and a minority of cell expressing the g/d heterodimer, which forms a similar structure. Dysregulation of dimerisation of membrane proteins can have deleterious effects on their function. In some cases, dimerisation can lead to the formation of inactive or misfolded protein complexes.

The dimerisation of membrane proteins can also have an impact on the properties of the membranes themselves. Membrane proteins are embedded in the lipid bilayer, which is composed of phospholipids and cholesterol. The presence of membrane proteins can affect the physical properties of the lipid bilayer, such as its fluidity, membrane lipid order, and thickness. This can in turn affect the function of other membrane proteins and the overall physiology of the cell.

4.2 FORSTER RESONANCE ENERGY TRANSFER (FRET)

4.2.1 Introduction to FRET

Forster (sometimes misnamed as fluorescence) resonance energy transfer (FRET) is a mechanism of energy transfer between two chromophores, where the excited state energy of one chromophore (the donor) is non-radiatively transferred to another chromophore (the acceptor) which is in close proximity (Forster 1946). FRET was first described by Theodor Forster in 1948, who developed a theoretical framework to describe the energy transfer process. FRET only occurs when the two chromophores, typically fluorescent molecules, are within a few (typically 1–10 nm) nanometres of each other (Figure 4.1) and their spectral properties are appropriately matched (there are requirements on the alignment of the electric dipoles and for overlap between the donor emission spectrum and the acceptor excitation spectrum). It can therefore be used as a short-range "molecular ruler" (Stryer 1978).

The energy transfer process involves three key steps: Firstly, the excitation of the donor chromophore by absorption of a photon – the same process as in standard fluorescence excitation. Secondly, non-radiative transfer of energy from the excited donor chromophore to the acceptor chromophore via dipole–dipole interaction, and finally, the emission of a photon by the acceptor chromophore – essentially the same process as standard fluorescence emission. This photon can be detected in the usual way using fluorescence microscopes including wide-field, TIRF, confocal, and other varieties (Jares-Erijman and Jovin 2006).

Because of the requirement for very close proximity between the donor and acceptor molecules, FRET has become an essential tool in modern biological research for investigating the interactions of biomolecules (Sekar

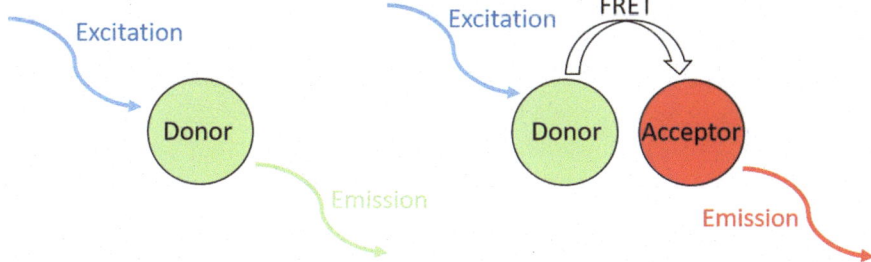

FIGURE 4.1   Principle of FRET. A conventional fluorophore (the donor) absorbs short wavelengths (blue), and due to the Stokes shift, emits longer wavelength fluorescence photons (green). If another fluorophore (the acceptor) is in close proximity (<10 nm) and its excitation spectrum overlaps with the donors emission spectrum, energy can be not radiatively transferred between the donor and the acceptor. The fluorescence emission will therefore be from the acceptor fluorophore and will be red-shifted compared to the donors original emission wavelengths.

and Periasamy 2003). It allows for the investigation of molecular interactions (e.g., protein–protein binding) and conformational changes in real-time (Miyawaki, Llopis et al. 1997). It should be noted, however, that FRET is not a method – it is a photophysical phenomenon. To detect FRET, there are a number of microscope technologies that can be employed – multi-colour, spectral imaging, fluorescence lifetime imaging microscopy (FLIM) and polarisation-resolved imaging (which is specifically used to detect homo-FRET; FRET between two identical fluorophores; Bader, Hofman et al. 2009). These different readouts have various advantages and disadvantages compared to each other, and the choice should be determined by the experimental sample and data requirements of the biological study. Additionally, each readout method can be implemented on a variety of microscope configurations including wide-field epi-fluorescence, confocal or multi-photon microscopy, TIRF microscopy, and even single-molecule or other super-resolution microscopies (Roy, Hohng et al. 2008).

4.2.2 Theoretical Basis

The equation for the inverse sixth power law for FRET, also known as the Forster distance, is:

$$E = \frac{1}{1 + (r/R_0)^6}$$

where E is the FRET efficiency (defined as the fraction of the donor energy that is non-radiatively transferred to the acceptor), R_0 is known as the Forster distance, which is the distance between the donor and acceptor molecules at which the FRET efficiency is 50% or 0.5. The value of R_0 therefore serves as a good "characteristic distance" that will be probed in a FRET experiment. r is the distance between the donor and acceptor molecules. As can be seen, the distance dependence is a 6th power, which means that the FRET efficiency drops off very quickly with distance, implying that FRET can only be used to measure distances below around 10 nm. Note that this places it in a different scale to the distances probed by SMLM where 10 nm is typically the resolution limit of that method (Lelek, Gyparaki et al. 2021).

The Forster distance is dependent on several factors, including the spectral properties of the donor and acceptor molecules, the orientation of their transition dipole moments, and the refractive index of the medium. The Forster distance can be calculated using the following equation:

$$R_0{}^6 \propto \frac{\kappa^2 Q_D}{n^4} J$$

where k^2 is the orientation factor, which describes the relative orientation of the transition dipole moments of the donor and acceptor molecules, Q_D is the quantum yield of the donor molecule, J is the spectral overlap integral, which describes the spectral overlap between the donor emission and acceptor absorption spectra, and n is the refractive index of the medium.

In many applications, especially in biological samples, measuring or even estimating these parameters can be difficult. For example, the value of the orientation factor is normally assumed to be 2/3. This is the value if the orientations of the donor and acceptor dipoles are essentially "randomised" which might be expected in molecules can rotate and re-orientate freely. If the motion of molecules is constrained, for example, by being embedded in a membrane or other large oligomeric stricture, this assumption might not be valid (Khrenova, Topol et al. 2015). The refractive index is often assumed to be 1.33 – the refractive index of water. However, the exact value in cells usually cannot be determined (Beuthan, Minet et al. 1996). Q_D and J can usually be measured using a spectrofluorometer, but again usually not in cells. Overall then, FRET is usually presented as a relative measure (i.e., FRET is increased or decreased between conditions) rather than attempting to calculate an actual intramolecular distance, which would require careful calibration.

4.2.3 Practical Considerations

One of the main challenges of measuring FRET in living cells is the need for appropriate fluorophores that can be used as FRET pairs (Bajar, Wang et al. 2016). The choice of fluorophores is critical as they must have suitable spectral properties to enable efficient energy transfer while also being compatible with cellular environments. For example, the use of CFP and YFP as FRET pairs is a common approach in live-cell imaging, but other pairs may be more appropriate depending on the specific experimental needs (Shaner, Steinbach et al. 2005). In addition, care must be taken to ensure that the fluorophores are expressed at appropriate levels and localised to the correct subcellular compartments to enable accurate FRET measurements.

Another important consideration is the use of appropriate instrumentation and imaging methods. FRET measurements require high-quality imaging systems that can accurately detect and quantify fluorescence signals. Confocal microscopy is commonly used for FRET imaging as it allows for high-resolution imaging of cells and provides excellent signal-to-noise ratios (Wallrabe, Elangovan et al. 2003). Other imaging modalities, such as total internal reflection fluorescence (TIRF) microscopy, can also be used for FRET imaging, depending on the specific experimental needs (Lin and Hoppe 2013).

A critical step in FRET measurement is the correction for spectral bleed-through and cross-talk. Spectral bleed-through occurs when the donor or acceptor fluorescence signal is detected in the emission channel of the other fluorophore. Cross-talk occurs when the acceptor fluorescence signal is excited directly by the donor excitation source, rather than by energy transfer. These effects can lead to false-positive or false-negative FRET signals if not properly corrected. Correction for spectral bleed-through and cross-talk can be performed using spectral unmixing algorithms or other analytical methods.

Finally, the use of appropriate controls is critical for accurate FRET measurements. Control experiments should be performed to verify that the observed FRET signals are not due to artefacts, such as nonspecific binding or photobleaching. Negative controls, such as cells expressing non-interacting proteins or cells expressing FRET pairs with a large separation distance, can be used to verify the specificity of the FRET signals. Positive controls, such as cells expressing FRET pairs with a known interaction or cells treated with drugs that affect protein–protein interactions, can be used to validate the FRET assay.

4.2.4 Intensity vs FLIM FRET

Spectral imaging is a widely used method for FRET measurement. In spectral imaging, the donor and acceptor fluorescence signals are separated into two detectors. While exciting the donor only, in the absence of FRET, the majority of the signal will appear in donor emission channel. When FRET occurs, the signal in the donor channel will decrease, and instead, signal will be detected in the acceptor detection channel (Figure 4.2). Spectral imaging has several advantages. Firstly, spectral imaging is a relatively simple and straightforward method that can be easily implemented using standard imaging systems. Secondly, spectral imaging provides high spatial resolution, allowing for the visualisation of FRET signals at the cellular and subcellular levels. The detection is also fact, allowing live-cell imaging studies or even video rate, or higher FRET measurements.

However, spectral imaging also has several limitations. One major limitation is the potential for spectral bleed-through and cross-talk, which can lead to false-positive or false-negative FRET signals. Correction for spectral bleed-through and cross-talk can be performed using spectral unmixing algorithms or other analytical methods, but these corrections can be difficult to apply accurately and may introduce additional errors. Calculation of FRET efficiencies can also be complicated by the unknown concentrations of the donor or acceptor, especially if fluorescent proteins are used and the expression levels cannot be well controlled.

FLIM is another method for measuring FRET that is based on the measurement of fluorescence lifetimes (Dowling, Dayel et al. 1998, Datta, Heaster et al. 2020). In FLIM, the fluorescence decay of the donor fluorophore is measured in the presence and absence of the acceptor fluorophore,

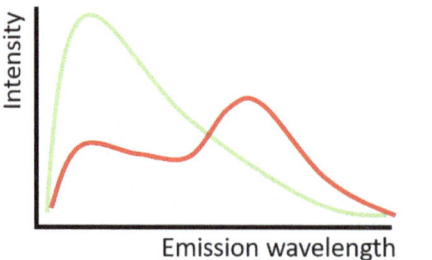
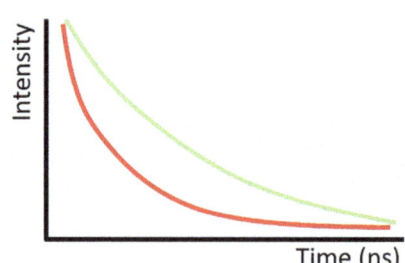

FIGURE 4.2 Two different methods for detecting FRET. Left: The green spectrum represents the emission observed in the absence of FRET, in which all emitted photons originate from the donor. Red represents the emission spectrum in the presence of FRET. Here, the donor emission is reduced and there is a new peak of emission intensity at longer wavelengths which originate from the acceptor.

and the change in fluorescence lifetime is used to calculate FRET efficiency. FRET provides an additional pathway for the donor to lose energy and therefore shortens the fluorescence lifetime (Figure 4.2). The higher the amount of FRET, the more the lifetime will be shortened. FLIM has several advantages over spectral imaging. Firstly, FLIM provides a more quantitative measurement of FRET efficiency, as it is less affected by spectral bleed-through or the requirements to know the concentrations of the molecules as it is an inherently ratiometric measurement. Secondly, FLIM can be used to measure subtle changes in protein–protein interactions or conformational changes that may not be detectable using other FRET methods. Finally, FLIM can be used to measure FRET in live cells over extended periods of time, allowing for the investigation of dynamic cellular processes.

However, FLIM also has several limitations. One major limitation is the requirement for specialised instrumentation and expertise. FLIM often requires the use of ultrafast laser systems and time-resolved detection systems, which can be expensive and technically challenging to implement. In addition, FLIM can be more time-consuming than other FRET methods as it requires the acquisition of multiple fluorescence decay curves for each FRET measurement. However, high-speed FLIM instrumentation has been and is being developed.

Finally, polarisation-resolved imaging can be used. This is a more specialised method and is typically used in special circumstances – notably when the donor and acceptor molecules are the same – homo-FRET, and therefore, no changes in the emission spectra or fluorescence lifetimes can be observed.

4.3 FRET IMAGING

4.3.1 Labelling Methods

In order to perform FRET measurements in cells, it is necessary to label the proteins of interest with appropriate fluorophores. Sample labelling strategies for FRET measurements in cells typically involve the use of genetically encoded fluorescent proteins or chemical dyes that can be attached to the proteins of interest using antibodies – immunostaining.

One commonly used strategy for FRET measurements in cells is to label the proteins of interest with genetically encoded fluorescent proteins such as green fluorescent protein (GFP), yellow fluorescent protein (YFP), cyan fluorescent protein (CFP), and red fluorescent protein (RFP) fused to the proteins of interest. The donor and acceptor fluorescent proteins can be chosen based on their spectral properties and the desired FRET efficiency. For example, CFP can be used as the donor and YFP as the acceptor as these fluorophores have a high overlap in their emission and excitation spectra, allowing for efficient energy transfer. There has been considerable research in developing fluorescent proteins with spectral properties conducive to FRET measurements (Shaner, Steinbach et al. 2005; Mastop, Bindels et al. 2017).

Another strategy for FRET measurements in cells is to use chemical dyes that can be attached to the proteins of interest. For example, the fluorescent dye Alexa Fluor 488 can be conjugated to a primary antibody targeting one protein, while the acceptor dye Alexa Fluor 555 used when labelling the other protein of interest (Chakraborty, Núñez et al. 2014). However, with this strategy, the sizes of the antibodies themselves can place a considerable separation between the dye molecules, resulting in a weak FRET signal.

It is important to note that sample labelling strategies for FRET measurements in cells must be carefully optimised to minimise artefacts and false-positive signals. For example, the labelling density must be carefully controlled to avoid nonspecific labelling or aggregation of the proteins of interest. As usual, the use of fluorescent fusion constructs can result in overexpression of the proteins of interest, misfolding or mislocalisation of the protein, or alterations to cell behaviour from the transfection process. It can therefore be difficult to ensure that FRET measurements accurately reflect the natural interactions and conformational changes of proteins in the cell.

The choice of fluorophores is critical for the success of FRET experiments, as the properties of the fluorophores greatly impact the efficiency and accuracy of FRET measurements. There are a number of factors that should be taken into account when selecting fluorophore pairs for FRET:

1. **Spectral overlap:** The donor and acceptor fluorophores should have a significant overlap between their emission and excitation spectra. This overlap ensures that the energy transfer between the donor and acceptor fluorophores is efficient and maximised.

2. **Brightness:** Fluorophores used in FRET experiments should be bright enough to allow for reliable detection. Brightness can be affected by factors such as the excitation wavelength and absorption

cross-section and the quantum yield of the fluorophore. Bright fluorophores lead to better signal-to-noise measurements and need less excitation intensity, resulting in lower photobleaching and lower phototoxicity.

3. **Photostability:** Fluorophores used in FRET experiments should be stable and resistant to photobleaching, which can lead to loss of signal and reduced accuracy of FRET measurements. The photostability of a fluorophore can be influenced by factors such as the chemical structure, the environment, and the excitation power.

4. **Minimal spectral cross-talk:** Fluorophores used in FRET experiments should have minimal spectral cross-talk, meaning that the emission spectra of the donor fluorophore should not overlap significantly with the excitation spectra of the acceptor fluorophore and vice versa. Spectral cross-talk can lead to false-positive FRET signals and reduced accuracy of FRET measurements.

5. For fluorescence lifetime measurements, fluorophores should have lifetimes that can be conveniently and accurately measured using the available instrumentation. Typically, this might involve natural fluorescence lifetimes in the 1.5–6 nm range before FRET.

4.3.2 Fluorescence Lifetime Imaging

FLIM is a technique that allows for the measurement of fluorescence lifetimes, which can provide information about the local environment and dynamics of fluorophores in biological samples. FLIM can be measured using a variety of methods, including time-correlated single photon counting (TCSPC), which is most commonly used (Becker, Bergmann et al. 2004), time-gated imaging, and frequency-domain FLIM. It is widely used to detect FRET (Wallrabe and Periasamy 2005), but it can also provide other readouts, for example, when combined with environmentally sensitive fluorophores (see Chapter 5; Owen, Lanigan et al. 2006).

TCSPC is a commonly used method for FLIM measurements (Becker, Bergmann et al. 2004). In TCSPC, a short laser pulse is used to excite the fluorophores, and the resulting fluorescence signal is detected by a photon-counting detector. The time delay between the excitation pulse and the detected photon is measured, and the fluorescence lifetime can be calculated from the distribution of these time delays. TCSPC is highly accurate and sensitive, but it can be time-consuming and require specialised equipment.

It used point detectors such as PMTs or APDs and so is the usual implementation on confocal or multi-photon microscope systems.

Time-gated imaging is another method for FLIM measurements. In time-gated imaging, a series of images are acquired with a time delay between the excitation pulse and the detection window using, in effect, a high-speed shutter called a gated optical intensifier (GOI; Grant, Zhang et al. 2008). The lifetime of the fluorophores can be calculated from the intensity decay of the fluorescence signal over time. Time-gated imaging can be faster than TCSPC and can provide spatial information about the fluorophores, but it can be less accurate and require careful optimisation of the gating parameters. However, it uses cameras and so is best suited to implementation on wide-field and TIRF-based microscopes.

Frequency-domain FLIM is another method for measuring fluorescence lifetimes (Verveer and Hanley 2009). In frequency-domain FLIM, the fluorophores are excited with a modulated light source, and the fluorescence signal is demodulated at a specific frequency. The phase shift and modulation of the fluorescence signal can be used to calculate the fluorescence lifetime. Frequency-domain FLIM can be fast and accurate, but it requires specialised equipment and can be affected by photobleaching and sample movement.

Another method for measuring fluorescence lifetimes is time-resolved fluorescence anisotropy imaging (TR-FAIM). In TR-FAIM, the anisotropy of the fluorescence signal is measured over time, providing information about the rotational diffusion of the fluorophores (Siegel, Suhling et al. 2003). The fluorescence lifetime can be calculated from the anisotropy decay curve. TR-FAIM can provide information about the orientation and mobility of the fluorophores, but it requires specialised equipment and can be affected by sample heterogeneity.

In addition to these methods, there are also hybrid techniques that combine FLIM with other imaging modalities, such as confocal microscopy or super-resolution microscopy. For example, stimulated emission depletion (STED) microscopy can be combined with FLIM to provide high-resolution images with fluorescence lifetime information (Szalai, Siarry et al. 2021). FLIM FRET can also be combined with high-throughput imaging, for example, for applications in drug discovery (Talbot, McGinty et al. 2008; Kumar, Alibhai et al. 2011).

Overall, the choice of FLIM measurement method depends on the specific research question, the properties of the fluorophores and sample, and the available equipment and resources. TCSPC is highly accurate and sensitive but can be time-consuming, while time-gated imaging and frequency-domain FLIM can provide faster measurements but may be less accurate. Hybrid techniques can provide additional spatial information and resolution, but they may require specialised equipment and expertise. Careful optimisation and validation of FLIM measurements are crucial for accurate and meaningful interpretation of the results.

4.3.3 Image Processing

The simplest approach to FRET analysis arises when spectral FRET measurements have been performed. Assuming that fluorescence has been collected in two well-separated spectral channels, the first goal is to measure the two intensities: Using image analysis software, select regions of interest (ROIs) in both the donor-only and FRET images and record the average pixel intensities. In order to obtain the donor-only image, it is usually possible to bleach the acceptor using high laser intensities at the acceptor fluorophore excitation wavelengths. The most basic calculation of the FRET efficiency is then

$$E = 1 - \frac{I_{DA}}{I_D}$$

where I_{DA} and I_D represent the intensity of the donor with and without the presence of the acceptor, respectively. It is important to account for any background fluorescence or autofluorescence by subtracting the corresponding background values from the measured intensities. It is also essential to validate the FRET measurements using appropriate positive and negative controls, for example, in cells transfected with the donor fluorophore only or in cells transfected with the donor covalently linked to the acceptor in close proximity to act as a positive control.

4.3.4 FLIM Fitting

The simplest model used for fitting fluorescence decay is the single-exponential decay model, which assumes a mono-exponential decay of the fluorescence intensity:

$$I(t) = Ae^{-t/\tau} + b$$

In this equation, $I(t)$ represents the fluorescence intensity at time t, A is the amplitude of the exponential decay, b is the background, and τ is the fluorescence lifetime. Nonlinear fitting algorithms, such as the Levenberg–Marquardt algorithm, can be employed to optimise the parameters A and τ to best fit the experimental data.

For samples containing multiple fluorophores or exhibiting more complex decay kinetics such as in the presence of FRET, a multiexponential decay model may be employed. This model assumes the presence of multiple exponential decay components in the fluorescence decay curve, each with its own amplitude and fluorescence lifetime. The decay curve is fitted by summing up the contributions from each exponential component. For example, for a double exponential:

$$I(t) = A_1 e^{-t/\tau_1} + A_2 e^{-t/\tau_2} + b$$

Phasor analysis is an alternative approach to extract fluorescence lifetimes from FLIM data that provides a graphical representation of fluorescence decay kinetics. Phasor analysis transforms the time-domain decay curve into a two-dimensional plot known as the phasor plot. The phasor plot provides a visual representation of the decay kinetics and allows for rapid determination of fluorescence lifetimes without the need for curve fitting.

The phasor plot is constructed using the Fourier transformation of the fluorescence decay curve (Figure 4.3). The real and imaginary components of the phasor plot represent the cosine and sine Fourier components, respectively. By converting the decay curve into phasor coordinates, each point in the phasor plot corresponds to a unique combination of fluorescence lifetime and relative amplitude (Ranjit, Malacrida et al. 2018).

The advantage of phasor analysis is its ability to rapidly classify the fluorescence decay kinetics of multiple fluorophores present in a sample. By pre-calculating the phasor coordinates corresponding to specific fluorescence lifetime values, it becomes possible to assign different regions in the phasor plot to different fluorescence lifetime components. This allows for quick and efficient determination of fluorescence lifetimes without the need for complex fitting procedures. Pixels of the image displaying single-exponential decays lie on a semicircle in phasor space and double exponentials lie inside the semicircle on a cord connecting the two individual lifetimes.

FLIM data can be visualised as a pseudo-coloured image to provide a qualitative representation of fluorescence lifetimes across a sample. This visualisation technique enhances the interpretation of FLIM data by assigning different colours to different fluorescence lifetime values. To

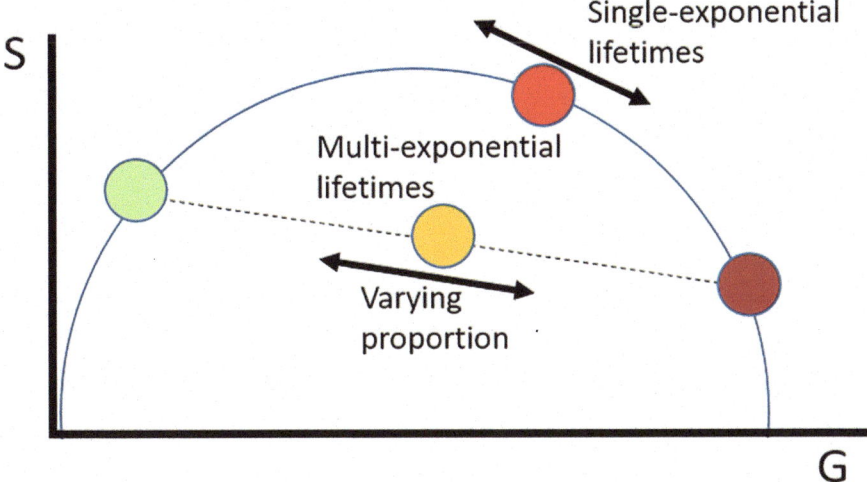

FIGURE 4.3 The phasor approach for analysing FLIM FRET data. By transforming each pixel's fluorescence decay curve into polar coordinates, the lifetimes can be visually inspected without the need for fitting. Single-exponential lifetimes lie on the semicircle, whereas multiexponential decays lie on a cord connecting the two individual decay components. As FRET increases, pixel positions will trace a path across phasor space.

create a pseudo-coloured image from FLIM data, a colour scale is established, where each colour corresponds to a specific fluorescence lifetime range. The colour scale is typically represented by a colour bar or a colour gradient. Shorter lifetimes are often associated with cooler colours such as blue or green, while longer lifetimes are represented by warmer colours such as yellow or red. Each pixel in the FLIM dataset is then assigned a colour based on its fluorescence lifetime value. For example, a pixel with a short lifetime would be assigned a blue colour, whereas a pixel with a longer lifetime would be assigned a red colour. Intermediate lifetime values can be represented by colours such as green, yellow, or orange. By representing fluorescence lifetimes as colours, the resulting pseudo-coloured image provides a visual representation of the spatial distribution of lifetimes within the sample. This enables the identification of regions with distinct fluorescence lifetime characteristics, highlighting variations in molecular interactions, probe environments, or physiological conditions across the sample.

Pseudo-coloured FLIM images can be further combined with intensity images to provide additional information about the spatial distribution of fluorophores or to correlate lifetime information with morphological features of the sample. This combined representation can facilitate the identification of specific regions of interest and enable the investigation of dynamic processes or molecular interactions within the sample. There are now several integrated FLIM packages for providing analysis and fitting for FLIM data, for example, Gao, Barber et al. (2020).

For FLIM data, the FRET efficiency is calculated as

$$E = 1 - \frac{\tau_{DA}}{\tau_A}$$

where *DA* represents the lifetime donor in the presence of the acceptor and *D* represents the lifetime of donor in the absence of the acceptor.

4.4 POTENTIAL ARTEFACTS AND BIOLOGICAL EXAMPLES

4.4.1 Artefacts

There are many potential artefacts that can affect FRET measurements. For example, heterogeneity of the sample: Biological samples often exhibit heterogeneity in terms of fluorophore distribution, probe localisation, or molecular interactions. This can lead to variations in FRET efficiency across different regions or subpopulations within the sample. It is important to account for this heterogeneity and perform appropriate statistical analysis or image segmentation techniques to obtain meaningful FRET measurements – for example, choosing appropriate ROIs from which to average lifetime or FRET values. In addition, the continuous exposure of fluorophores to excitation light during FRET measurements can lead to photobleaching, where the fluorescence intensity gradually decreases over time. Photobleaching can result in changes in FRET efficiency because the donor and acceptor molecules will be affected differently, changing their ratio. It is crucial to monitor and correct for photobleaching effects during FRET experiments using appropriate controls or photostable fluorophores. In an additional complication, variations in the photophysical properties of fluorophores, such as their quantum yields, fluorescence lifetimes, or photostability, can introduce artefacts in FRET measurements. These variations can lead to differences in the apparent FRET efficiency or affect the accuracy of lifetime-based FRET measurements. It is important to carefully choose fluorophores with well-characterised properties and account for any variations or discrepancies in the experimental setup. For example, many fluorophores show an environmentally dependent fluorescence lifetime, with lifetimes varying with parameters such as refractive index, pH, solvent polarity, viscosity, or other parameters. All of these bio-physical properties can vary within a cell.

The imaging setup can also lead to a number of potential artefacts. One of the most obvious is fluorescence cross-talk. Fluorescence cross-talk occurs when the emitted fluorescence from the donor or acceptor fluoro-phore is detected by the opposite channel due to spectral overlap or bleed-through. This can result in false-positive FRET signals or overestimation of FRET efficiency. To minimise cross-talk artefacts, appropriate spectral filters and correction algorithms should be employed, and control experi-ments without FRET should be performed to establish baseline measure-ments. Autofluorescence is another source of signal which can interfere with calculating FRET. Biological samples can exhibit inherent autofluo-rescence, which can interfere with FRET measurements. Autofluorescence signals may overlap with the fluorescence emission of the donor or accep-tor fluorophores, leading to false-positive FRET signals or inaccurate FRET efficiency calculations. It is crucial to characterise and subtract autofluorescence signals using appropriate controls or spectral unmixing techniques to obtain reliable FRET measurements. Autofluorescence var-ies with different excitation and emission wavelengths and can originate in a number of biological molecules from NAD to collagen.

Further artefacts can arise as a result of the fitting or data analysis pro-cedures, for example, due to inappropriate model assumptions. Fitting FRET data often involves assumptions about the FRET system, such as a simple two-state FRET model. However, complex FRET systems may devi-ate from these assumptions, leading to inaccurate results. For instance, multiple FRET pathways, heterogeneity in donor–acceptor distances, or non-ideal dye orientation can impact the FRET efficiency. It is crucial to consider alternative models and perform model selection or model-free analysis to avoid bias in FRET measurements. Limited data quality or resolution can also have an impact. Noisy or low-signal data can result in inaccurate fitting and lead to biased FRET efficiency calculations. It is important to optimise experimental conditions, acquire a sufficient num-ber of photons, and employ appropriate signal-to-noise ratio enhancement techniques to improve data quality. Fitting FRET data involves estimat-ing various parameters, such as fluorescence lifetimes, spectral overlap, or FRET efficiency. Correlation between parameters or overfitting can occur when the number of fitted parameters is too high relative to the available data. This can lead to unrealistic or unreliable parameter estimates and inaccurate FRET efficiency calculations. Regularisation techniques, model simplification, or cross-validation methods can help mitigate this issue. Biases can also occur during data analysis, such as the selection of ROIs or subjective thresholding. Inconsistent or biased ROI selection can lead to biased FRET efficiency measurements, while subjective thresholding can introduce variability in data analysis. Employing automated or objective analysis methods and ensuring consistent analysis criteria can minimise these biases.

4.4.2 Biological Examples

A good example of FLIM FRET being used to study the distribution of membrane proteins is the study of Ras (Miyawaki, Llopis et al. 1997). Ras is expressed in three isoforms (K-Ras, N-Ras, and H-Ras) each with varying post-translational lipid modification. Ras is one of the proteins most commonly mutated in cancers. Since the distribution of membrane proteins affects their signalling dynamics, the nanoscale organisation of Ras is an important research topic. FLIM FRET was able to detect differences in the nanoscale clustering of proteins between the three Ras isoforms. As a second example, single-molecule FRET has been used to analyse the dimerisation of GPCRs on the cell surface. mGluR2, a glutamate receptor which is expressed exclusively in the brain, forms stable dimers regardless of the density of receptors on the cell surface. Conversely, SecR, the secretin receptor, must be present at a surface density high enough to establish relatively long-lived interactions. Finally, MOR, the m-opioid receptor, is monomeric, regardless of surface density, and shows no evidence of forming dimers or any high-level oligomers (Asher, Geggier et al. 2021).

Overall, when membrane proteins present as small oligomers such as dimers, FRET is often the method of choice for detecting and analysing those complexes. FRET occurs only when molecules are within a few nanometres of each other and FLIM provides one of the most robust measurements of this process. FLIM can be implemented on a variety of microscope systems including confocal and TIRF systems to provide a powerful tool for analysing membrane protein dimerisation.

REFERENCES

Asher, W. B., P. Geggier, M. D. Holsey, G. T. Gilmore, A. K. Pati, J. Meszaros, D. S. Terry, S. Mathiasen, M. J. Kaliszewski, M. D. McCauley, A. Govindaraju, Z. Zhou, K. G. Harikumar, K. Jaqaman, L. J. Miller, A. W. Smith, S. C. Blanchard and J. A. Javitch (2021). "Single-molecule FRET imaging of GPCR dimers in living cells." *Nature Methods* **18**(4): 397–405.

Bader, A. N., E. G. Hofman, J. Voortman, P. M. en Henegouwen and H. C. Gerritsen (2009). "Homo-FRET imaging enables quantification of protein cluster sizes with subcellular resolution." *Biophysical Journal* **97**(9): 2613–2622.

Bajar, B. T., E. S. Wang, S. Zhang, M. Z. Lin and J. Chu (2016). "A guide to fluorescent protein FRET Pairs." *Sensors (Basel)* **16**(9): 1488.

Becker, W., A. Bergmann, M. A. Hink, K. König, K. Benndorf and C. Biskup (2004). "Fluorescence lifetime imaging by time-correlated single-photon counting." *Microscopy Research and Technique* **63**(1): 58–66.

Beuthan, J., O. Minet, J. Helfmann, M. Herrig and G. Müller (1996). "The spatial variation of the refractive index in biological cells." *Physics in Medicine and Biology* **41**(3): 369–382.

Chakraborty, S., D. Núñez, S.-Y. Hu, M. P. Domingo, J. Pardo, A. Karmenyan, G. Eva Ma and A. Chiou (2014). "FRET based quantification and screening technology platform for the interactions of Leukocyte Function-Associated Antigen-1 (LFA-1) with InterCellular Adhesion Molecule-1 (ICAM-1)." *PLOS One* **9**(7): e102572.

Clatot, J., M. Hoshi, X. Wan, H. Liu, A. Jain, K. Shinlapawittayatorn, C. Marionneau, E. Ficker, T. Ha and I. Deschênes (2017). "Voltage-gated sodium channels assemble and gate as dimers." *Nature Communications* **8**(1): 2077.

Datta, R., T. M. Heaster, J. T. Sharick, A. A. Gillette and M. C. Skala (2020). "Fluorescence lifetime imaging microscopy: fundamentals and advances in instrumentation, analysis, and applications." *Journal of Biomedical Optics* **25**(7): 1–43.

Dowling, K., M. J. Dayel, M. J. Lever, P. M. W. French, J. D. Hares and A. K. L. Dymoke-Bradshaw (1998). "Fluorescence lifetime imaging with picosecond resolution for biomedical applications." *Optics Letters* **23**(10): 810–812.

Forster, T. (1946). "Energiewanderung und Fluoreszenz." *Naturwissenschaften* **33**(6): 166–175.

Gao, D., P. R. Barber, J. V. Chacko, M. A. Kader Sagar, C. T. Rueden, A. R. Grislis, M. C. Hiner and K. W. Eliceiri (2020). "FLIMJ: an open-source ImageJ toolkit for fluorescence lifetime image data analysis." *PLoS One* **15**(12): e0238327.

Grant, D. M., W. Zhang, E. J. McGhee, T. D. Bunney, C. B. Talbot, S. Kumar, I. Munro, C. Dunsby, M. A. Neil, M. Katan and P. M. French (2008). "Multiplexed FRET to image multiple signaling events in live cells." *Biophysical Journal* **95**(10): L69–71.

Jares-Erijman, E. A. and T. M. Jovin (2006). "Imaging molecular interactions in living cells by FRET microscopy." *Current Opinion in Chemical Biology* **10**(5): 409–416.

Khrenova, M., I. Topol, J. Collins and A. Nemukhin (2015). "Estimating orientation factors in the FRET theory of fluorescent proteins: the TagRFP-KFP pair and beyond." *Biophysical Journal* **108**(1): 126–132.

Kumar, S., D. Alibhai, A. Margineanu, R. Laine, G. Kennedy, J. McGinty, S. Warren, D. Kelly, Y. Alexandrov, I. Munro, C. Talbot, D. W. Stuckey, C. Kimberly, B. Viellerobe, F. Lacombe, E. W. F. Lam, H. Taylor, M. J. Dallman, G. Stamp, E. J. Murray, F. Stuhmeier, A. Sardini, M. Katan, D. S. Elson, D. S. Elson, M. A. A. Neil, C. Dunsby and P. M. W. French (2011). "FLIM FRET technology for drug discovery: automated multiwell-plate high-content analysis, multiplexed readouts and application in situ." *Chemphyschem: A European Journal of Chemical Physics and Physical Chemistry* **12**(3): 609–626.

Lelek, M., M. T. Gyparaki, G. Beliu, F. Schueder, J. Griffié, S. Manley, R. Jungmann, M. Sauer, M. Lakadamyali and C. Zimmer (2021). "Single-molecule localization microscopy." *Nature Reviews Methods Primers* **1**: 1–27.

Lin, J. and A. Hoppe (2013). "Uniform total internal reflection fluorescence illumination enables live cell fluorescence resonance energy transfer microscopy." *Microscopy and Microanalysis: The Official Journal of Microscopy Society of America, Microbeam Analysis Society, Microscopical Society of Canada* **19**: 1–10.

Mastop, M., D. S. Bindels, N. C. Shaner, M. Postma, T. W. J. Gadella and J. Goedhart (2017). "Characterization of a spectrally diverse set of fluorescent proteins as FRET acceptors for mTurquoise2." *Scientific Reports* **7**(1): 11999.

Miyawaki, A., J. Llopis, R. Heim, J. M. McCaffery, J. A. Adams, M. Ikura and R. Y. Tsien (1997). "Fluorescent indicators for Ca2+based on green fluorescent proteins and calmodulin." *Nature* **388**(6645): 882–887.

Orloff, D. G., C. S. Ra, S. J. Frank, R. D. Klausner and J. P. Kinet (1990). "Family of disulphide-linked dimers containing the zeta and eta chains of the T-cell receptor and the gamma chain of Fc receptors." *Nature* **347**(6289): 189–191.

Owen, D. M., P. M. P. Lanigan, C. Dunsby, I. Munro, D. Grant, M. A. A. Neil, P. M. W. French and A. I. Magee (2006). "Fluorescence lifetime imaging provides enhanced contrast when imaging the phase-sensitive dye di-4-ANEPPDHQ in model membranes and live cells." *Biophysical Journal* **90**(11): L80–L82.

Owen, D. M., D. Williamson, C. Rentero and K. Gaus (2009). "Quantitative microscopy: protein dynamics and membrane organisation." *Traffic* **10**(8): 962–971.

Ranjit, S., L. Malacrida, D. M. Jameson and E. Gratton (2018). "Fit-free analysis of fluorescence lifetime imaging data using the phasor approach." *Nature Protocols* **13**(9): 1979–2004.

Roy, R., S. Hohng and T. Ha (2008). "A practical guide to single-molecule FRET." *Nature Methods* **5**(6): 507–516.

Sekar, R. B. and A. Periasamy (2003). "Fluorescence resonance energy transfer (FRET) microscopy imaging of live cell protein localizations." *Journal of Cell Biology* **160**(5): 629–633.

Shaner, N. C., P. A. Steinbach and R. Y. Tsien (2005). "A guide to choosing fluorescent proteins." *Nature Methods* **2**(12): 905–909.

Siegel, J., K. Suhling, S. Lévêque-Fort, S. E. D. Webb, D. M. Davis, D. Phillips, Y. Sabharwal and P. M. W. French (2003). "Wide-field time-resolved fluorescence anisotropy imaging (TR-FAIM): imaging the rotational mobility of a fluorophore." *Review of Scientific Instruments* **74**(1): 182–192.

Stryer, L. (1978). "Fluorescence energy transfer as a spectroscopic ruler." *Annual Review of Biochemistry* **47**(1): 819–846.

Szalai, A. M., B. Siarry, J. Lukin, S. Giusti, N. Unsain, A. Cáceres, F. Steiner, P. Tinnefeld, D. Refojo, T. M. Jovin and F. D. Stefani (2021). "Super-resolution imaging of energy transfer by intensity-based STED-FRET." *Nano Letters* **21**(5): 2296–2303.

Talbot, C. B., J. McGinty, D. M. Grant, E. J. McGhee, D. M. Owen, W. Zhang, T. D. Bunney, I. Munro, B. Isherwood, R. Eagle, A. Hargreaves, C. Dunsby, M. A. A. Neil and P. M. W. French (2008). "High speed unsupervised fluorescence lifetime imaging confocal multiwell plate reader for high content analysis." *Journal of Biophotonics* **1**(6): 514–521.

Verveer, P. J. and Q. S. Hanley (2009). Chapter 2 frequency domain FLIM theory, instrumentation, and data analysis. *Laboratory Techniques in Biochemistry and Molecular Biology* **33**: 59–94.

Wallrabe, H., M. Elangovan, A. Burchard, A. Periasamy and M. Barroso (2003). "Confocal FRET microscopy to measure clustering of ligand-receptor complexes in endocytic membranes." *Biophysical Journal* **85**(1): 559–571.

Wallrabe, H. and A. Periasamy (2005). "Imaging protein molecules using FRET and FLIM microscopy." *Current Opinion in Biotechnology* **16**(1): 19–27.

Wang, J.-H., K. Lim, A. Smolyar, M.-K. Teng, J.-H. Liu, A. G. D. Tse, J. Liu, R. E. Hussey, Y. Chishti, C. T. Thomson, R. M. Sweet, S. G. Nathenson, H.-C. Chang, J. C. Sacchettini and E. L. Reinherz (1998). "Atomic structure of an αβ T cell receptor (TCR) heterodimer in complex with an anti-TCR fab fragment derived from a mitogenic antibody." *The EMBO Journal* **17**(1): 10–26.

CHAPTER 5

Imaging Membrane Biophysical Properties

5.1 AN INTRODUCTION TO MEMBRANE BIOPHYSICAL PROPERTIES

Membranes are complex structures compartmentalising processes in the cells to sustain life. Membranes have a diverse biochemical composition which varies on all levels: between the organisms, tissues, cell types, and organelles. The main structural components of cellular membranes are proteins and lipids, where lipids form a bilayer and proteins are either inserted in the bilayer or are associated with it. The amphipathic nature of lipids allows them to assemble in a highly stable structure where hydrophobic parts of lipids are oriented to each other in both leaflets forming the core of the bilayer with hydrophilic groups exposed to the exterior of the membrane. This model of cellular membrane is called the fluid mosaic model and has been proposed by Singer and Nicolson (1972).

For a long time, proteins have been regarded as the main players in defining the biological function of membranes, and lipids were mainly providing the medium for the proteins and the barrier function for the membranes. Currently, this concept is challenged suggesting that membrane lipid composition plays a critical role in various cellular processes such as cellular proliferation (Preta 2020), division (Atilla-Gokcumen, Muro et al. 2014), differentiation (Barceló-Coblijn, Martin et al. 2011), signalling (Sunshine and Iruela-Arispe 2017), and mechanosensation (Pliotas, Dahl

 DOI: 10.1201/9781003273745-5

et al. 2015) through either affecting biophysical properties of the cell membrane or/and signalling that occurs on the membrane.

Lipids and particularly membrane lipids are one the most chemically diverse group of biomolecules (Harayama and Riezman 2018). With hundreds of different species that mammalian cells are generating, the main property that unites membrane lipids is their amphipathic nature, which means that the same molecule will share both hydrophobic and hydrophilic parts. The hydrophobic part of the molecule is directed to the corresponding hydrophobic part of the opposing leaflet of the membrane and the hydrophilic headgroup is exposed to the surroundings of the membrane. The major classes of membrane lipids are glycerophospholipids, sphingolipids, and sterols. Chemical diversity in glycerophospholipid and sphingolipid classes is generated by combining various headgroups to either glycerol or sphingosine backbone. Also, the hydrophobic part of these molecules varies greatly by combining various fatty acid moieties to the backbones which can be of various carbon atom lengths and contain double bonds or hydroxyl groups. There is also a huge chemical diversity in sterols, some of which define phylogenetic differences in organisms; for instance, animals, fungi, and plants synthesise various types of sterols (Desmond and Gribaldo 2009).

A lipid bilayer can be composed of a single lipid type and even complex lipid behaviour can be reconstituted in three component model lipid membranes. Therefore, an obvious question is why organisms invest energy in generating vast chemical diversity of membrane lipids. One possibility is that lipids define the biophysical properties of the membrane and therefore are the subject of complex homeostatic regulation. Chemical diversity in lipids allows cells to finely tune these properties in response to the environment, nutrient availability, and metabolic context in which they operate.

The biophysical properties of membranes are critical on both macro- and nanoscale. In cells, membranes are highly dynamic structures forming various morphologies that are important for biological processes. Bending membranes, forming vesicles of various sizes, adhesions, and dividing cells, require flexibility that is regulated by lipid composition.

Moreover, nanoscale properties of membranes also affect protein function on the membrane. Around a third of the protein in cells are transmembrane meaning membranes (Krogh, Larsson et al. 2001), and specifically, the local environment that lipids generate around transmembrane proteins is critical for a substantial number of processes in the cells (Levental and

Lyman 2023). The interrelationship between proteins and lipids in membranes is complex and ultimately results in collective membrane biophysical properties. A canonical example of such interaction is that membrane organisation affects protein diffusion and dynamics (Owen, Williamson et al. 2012). In addition, the hydrophobic thickness of the lipid bilayer distributes proteins based on the length of the transmembrane domains of the proteins (Andersen and Koeppe 2007). The mechanical properties of the membranes also influence ion channels, regulating the permeability of the membrane (Andersen and Koeppe 2007).

Because of chemical diversity, some lipids prefer to interact with certain lipids and repulse others. This basic understanding explains why lipids segregate in simple and complex mixtures to form nanodomains, also called "lipid rafts" (Simons and Ikonen 1997) (Figure 5.1). In physiologically relevant conditions, particularly interesting is liquid–liquid phase separation which is characterised by the coexistence of two liquid phases with dissimilar biophysical properties. In various conditions, such liquid–liquid phase separation has been observed in synthetic or biomimetic membranes: when

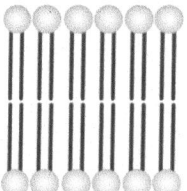

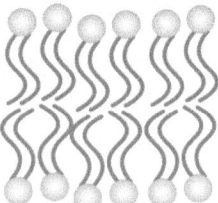

The gel phase   The liquid ordered phase   The liquid disordered phase

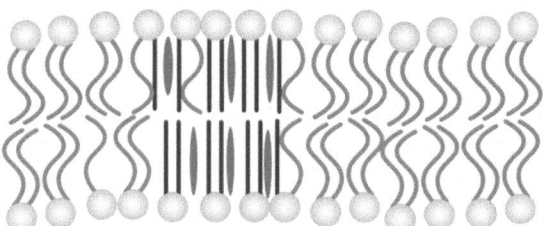

- **Lipid packing**
- **Fluidity**
- **Viscosity**
- **Tension**

FIGURE 5.1 Lipid bilayers can exist in one of three phases – the gel phase, the liquid-ordered phase, and the liquid-disordered phase depending on the lipid composition. The liquid-ordered phases are often termed lipid rafts and are enriched in saturated phospholipids and cholesterol. In complex compositions such as cell plasma membranes, lipids (and proteins) can segregate creating coexisting phases within the bilayer.

mixing saturated lipids (which lack double bonds and form straight rigid acyl tails) and unsaturated lipids (which have double bonds making them more flexible) with sterols can separate into two liquid phases. Lipids with saturated acyl tails prefer to interact with sterols and form liquid-ordered (Lo) phase, which is characterised by highly ordered packing, higher stiffness, and thickness of bilayer compared to liquid-disordered (Ld) phase that is composed of lipids with unsaturated acyl tails.

The lateral organisation of lipids in membranes drives the partitioning of the proteins that would prefer distinct membrane environments. This self-organisation generates conditions in which proteins and therefore function associated with them are influenced by lipid composition. Collective biophysical membrane properties such as surface potential, lipid packing, tension, fluidity, and intrinsic curvature are defined by the chemical diversity of lipid components of the membrane and can be measured by various methods. Here, we will discuss how using small molecules called probes we can measure various biophysical properties. How we can image and analyse the data and how we can apply these methods to answer questions in biology.

5.2 ENVIRONMENTALLY SENSITIVE PROBES

5.2.1 Introduction

Cellular membranes and the diversity of physical properties are investigated using fluorogenic small molecules that have capacity to penetrate membrane lipid bilayer. These molecules, known as environmentally sensitive dyes, possess the ability to modulate their emission characteristics in response to intrinsic features of their surrounding molecular environment. Due to their small size and ability to be modified, these probes are considered to have minimal impact on cell physiology and can be tagged to specific cellular organelles. Various mechanisms are involved in changing emission properties of environmentally sensitive dyes, but we will focus here on their application and division based on the specific properties that these dyes can report.

5.2.2 Probe Types and Mechanisms

How lipids are organised in the membranes and how they are packed is critical for the overall biophysical characterisation of membranes. This information can be obtained directly from the fluorescent images using small probes and fluorescent microscopy.

One of the main types of probes which can report on lipid packing is one that undergoes a large change in their dipole moment upon excitation. These dyes are aromatic structures that are able to transfer an electron from donor to acceptor group within the molecule. In the ground state, polar molecules in the solvent (usually water) are arranged around the dye dipole in a low energy configuration. With the absorption of a photon and excitation of the dye, the dipole changes dramatically and almost instantaneously. This leaves the solvent molecules in a high energy configuration leading to a high energy excited state (blue emission). Over time – several nanoseconds, the solvent molecules rearrange and adopt a low energy configuration again. This lowers the energy of the excited state and causes a red-shift of the emission. The more polar the surrounding solvent, the more the molecules can interact with the dye and the more pronounced this process will be. This means the emission wavelength of the fluorophore becomes red-shifted with increasing polarity of the dye's solvent. These probes are therefore called the polarity-sensitive probe which in the context of lipid bilayer sense the amount of water penetrating into the hydrophobic core of the membrane, which is where the dye's chromophore is typically located. If the lipid bilayer is tightly packed, there will be fewer spaces for water molecules to penetrate and so the probe will experience fewer polar dipoles surrounding it. This will result in a different fluorescence signal than if the lipid bilayer is more loosely packed, allowing more water molecules to penetrate and surround the probe (Figure 5.2). By measuring changes in their emission spectra, we can identify the lipid packing in the membrane (Bagatolli 2006).

The probes that can report on lipid order can be divided based on their spectroscopic characteristics and their permeability and cellular localisation. Spectroscopic features, such as excitation wave, emission spectra, and photostability, are critical especially if working with live-cell sample or monitoring the dynamics of these properties. One of the first designed polarity-sensitive dyes that measure lipid packing is Laurdan (Parasassi, De Stasio et al. 1991). Laurdan (1-[6-(dimethylamino)-2-naphthalenyl]-1-dodecanone) is a commercially available probe that can be used with confocal microscopes if certain parameters of excitation and collection of emission are in place. This probe is excited by UV (around 340 nm) and emission spectra which depends on the lipid packing: 440 nm for liquid-ordered phase and 490 nm for liquid-disordered phase. Other characteristics that make Laurdan suitable for measuring lipid order in the membranes of living cells are: it homogeneously distributes between

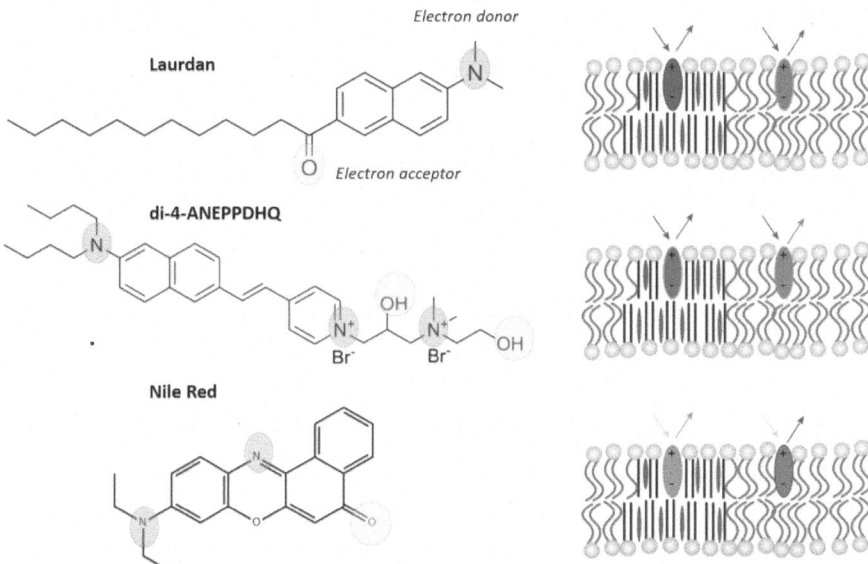

FIGURE 5.2   Schematics of the chemical structures of the environmentally sensitive dyes molecules Laurdan: di-4-ANEPPDHQ and Nile Red. All of these can insert into the lipid bilayer and change their fluorescent properties depending on the phase behaviour (membrane lipid packing) of the surrounding molecules.

phases without discriminating for liquid-ordered or -disordered phase, it is not soluble in the water; therefore, minimising background for imaging, its emission spectrum is shifted to blue and green, allowing to perform simultaneous microscopy with red emitting fluorophores.

Despite being one of the most known polarity-sensitive dyes reporting on the lipid packing, Laurdan has its own limitations: firstly, excitation with UV is photodamaging for the live cells, and secondly, the dye has high photobleaching which restricts it from being used in timelapse experiment setup. Initially developed as a voltage sensing probe, another similarly working dye di-4-ANEPPDHQ is used to image lipid packing (Jin, Millard et al. 2006). Di-4-ANEPPDHQ excites at 488 nm and has a relatively red-shifted spectrum with emission maxima of 560 and 610 nm in ordered and disordered liquid phase membranes, respectively. This optical setup is more common, and general confocal system can be used to image lipid order with di-4-ANEPPDHQ. Also, live-cell imaging with excitation of 488 nm in the case of di-4-ANEPPDHQ is more gentle and allows to perform timelapses compared to Laurdan that excites at UV and has detrimental effect on live cells. Di-4-ANEPPDHQ can even be used in whole living organisms such as zebrafish (Owen, Magenau et al. 2010; Owen, Rentero et al. 2012) and can also be used with fluorescence lifetime imaging as the readout (see Chapter 4; Owen, Lanigan et al. 2006). A large variety of similar probes are now available. These sense membrane properties in the same way, but have a range of standard fluorescence properties such as excitation and emission spectra (Kwiatek, Owen et al. 2013).

Laurdan and di-4-ANEPPDHQ exhibit different reactions to solvents of varying polarity, indicating distinct mechanisms behind their spectral shifts. Laurdan fluorescence is highly sensitive to temperature, while di-4-ANEPPDHQ fluorescence is responsive to cholesterol content in liposomes. The time-dependent fluorescence shifts reveal that di-4-ANEPPDHQ behaves differently from Laurdan, suggesting a non-dipolar relaxation pattern. These dyes reflect different membrane features, and their fluorescence spectra are influenced by hydration, mobility, cholesterol content, and membrane potential. These differences should be considered when using these dyes for cell biology studies involving multiple processes (Amaro, Reina et al. 2017).

Solvatochromic dyes provide a valuable tool for investigating lipid order at the subcellular level within various organelles. Notably, organelle-targeted probes have been created using Nile Red, another solvatochromic dye. These probes enable the imaging of lipid order in specific organelles, expanding our ability to study subcellular lipid organisation. Nile Red, initially used as a lipid droplet marker, offers several advantages such as high brightness, operation in the optimal spectral range (yellow-red), and sensitivity to lipid order. Its fluorogenic properties have enabled spectrally resolved super-resolution imaging of lipid order. By incorporating membrane anchor groups, Nile Red derivatives (NR12S, NR12A, and NR4A) have been developed for ratiometric and spectral imaging of lipid order specifically in plasma membranes. Thus, the functionalisation of Nile Red with organelle-targeting groups holds great potential for studying polarity and lipid order in various organelles within live cells (Danylchuk, Jouard et al. 2021).

5.2.3 Tension Probes

Another critical biophysical property of the membrane is its stiffness, rigidity, and tension which can also be measured using small molecule reporters. Membrane tension is vital in multiple cellular processes. It is regulated during cell migration, spreading, phagocytosis, and division. Additionally, it influences endocytosis, ion channel opening, and cell metabolism. Similar to measuring lipid packing, "push and pull" can be used to measure mechanistic properties of the membranes simply because membrane tension is expected to alter lipid packing. However, even dramatic changes in tension only slightly change lipid packing, therefore probes that are extremely sensitive can be used to measure tension changes.

One of such probes has been developed and is called FliptR (for "fluorescent lipid tension reporter"; Colom, Derivery et al. 2018). Two fluorescent dithienothiophene groups called flippers are connected by a rotatable bond. The positions of the two fluorescent groups along the axis can be influenced by the pressure exerted by the surrounding lipids (Figure 5.3).

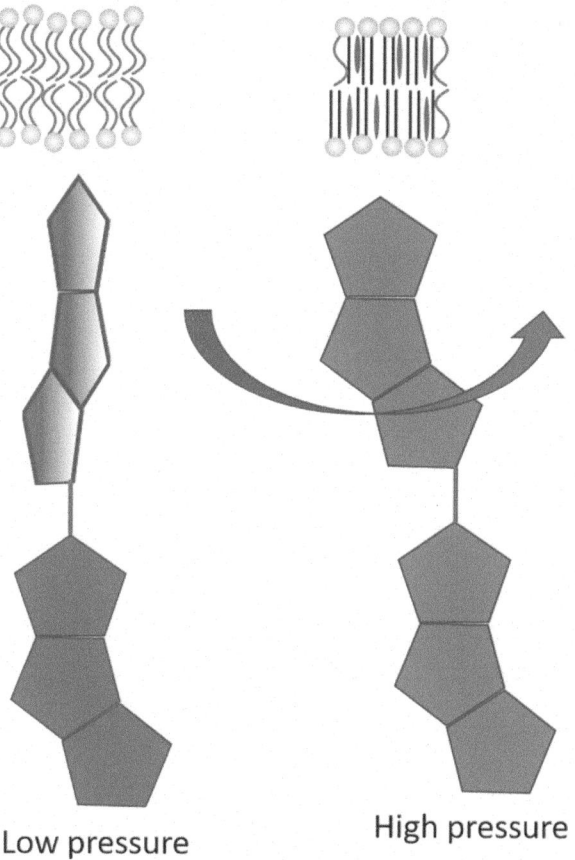

Low pressure   High pressure

FIGURE 5.3   Example of the mechanism of action of a membrane pressure sensor. Changes in pressure cause a rotation within the molecule causing a change in the alignment of two aromatic structural elements. This manifests in a change in the molecules fluorescent properties which can be read out in a fluorescence microscope.

When these two groups are oriented perpendicular to each other, they experience low tension, which is indicative of lower acyl tail packing density in the lipid bilayer. Conversely, high pressure and tension changes the orientation of these two fluorescent so that they are situated in the same plane.

The relative position of two fluorescent groups affects the fluorescence lifetime of the probe. Using fluorescence lifetime imaging microscopy (FLIM) and FliptR provides an effective method for imaging membrane tension in live cells. It possesses several key characteristics that make it ideal for this purpose. Firstly, it partitions into different membrane domains without a strong preference. Secondly, it causes minimal disturbance to the surrounding membrane structure. Thirdly, it specifically labels the plasma membrane. Fourthly, it exhibits negligible fluorescence in water. Finally, it is insensitive to membrane potential.

While FliptR has a limitation in that it reports lipid packing, which is influenced by both tension and lipid composition, it is challenging to conceive an alternative method for sensing membrane tension at the molecular level without considering lipid packing. Therefore, developing a membrane tension probe that is unaffected by lipid composition is unlikely. Despite this limitation, FliptR reliably detects changes in membrane tension, which occur within a few seconds and can be easily distinguished from lipid composition changes that take place over tens of minutes.

5.2.4 Tension Probes

Viscosity and the way molecules diffuse in membranes are extremely important biological parameters of the membranes. Another class of small molecules are molecular rotors which are capable of detecting changes in viscosity through alterations in their fluorescence intensity or lifetime (Kuimova, Yahioglu et al. 2008; Kuimova 2012).

These molecular rotors exhibit different decays of the excited state in response to the viscosity of their immediate micro-environment. This sensitivity enables accurate calibration of fluorescence parameters in relation to viscosity. In the same way as viscosity probes, molecular rotor requires a combination of FLIM and ratiometric imaging analysis.

Similar to lipid order probes, numerous molecular rotors have been developed for the purpose of imaging viscosity within internal cellular organelles, including specific probes for lysosomes, mitochondria, genetically targeted probes, and recently, to specific rotor for the plasma membrane (López-Duarte, Vu et al. 2014).

5.3 IMAGING

5.3.1 Dye Selection

When selecting a probe for studying membrane properties, several criteria should be considered. One important factor is the availability probes. Examples of widely used commercially available probes include Laurdan and di-4-ANEPPDHQ for imaging lipid order, as well as Flipper-TR for investigating membrane tension. However, it is worth noting that not all probes may be commercially available. Some probes might be more challenging to obtain for experimental purposes, as they may not be readily accessible through commercial sources. In such cases, researchers may need to explore alternative approaches to acquire the specific probes they require for their studies.

Probes should exhibit low background fluorescence in the polar and aqueous environment surrounding the membrane. This is crucial to minimise interference and noise during measurements.

Also, the probes that are designed to measure lipid order should demonstrate a significant difference in Stokes shift between the ordered and disordered phases of the membrane. The Stokes shift refers to the difference in wavelengths between the absorption and emission of light by the probe. A large difference in Stokes shift allows for a more pronounced change in the emission spectra of the probe in response to incremental alterations in membrane order. This ensures a greater sensitivity and dynamic range for detecting variations in membrane order.

Photostability is another important criteria one should consider when choosing the suitable probe especially in live-cell imaging for time lapse experimental setup. For instance, di-4-ANEPPDHQ is more photostable compared to Laurdan and therefore more suitable for the timelapse imaging. Internalisation into different membrane compartments can be an obstacle if one desires to image specific organelle since different probes used for studying membranes exhibit variations in their distribution and dynamics across various membrane compartments. For live-cell imaging, cell toxicity of chosen probe can also be an obstacle in imaging membrane properties.

Measuring tension or viscosity of the membranes in live or fixed cells can be done, but it is more technically challenging. These probes often rely on fluorescence lifetime imaging, and in some cases, they may also utilise ratiometric imaging. The key consideration when using these probes is their spectroscopic properties, as some probes are not suitable for live-cell imaging due to their excitation/emission spectra being shifted towards the UV range. Probes with excitation/emission spectra in the visible or near-infrared range are commonly preferred for live-cell applications, as they offer better penetration and reduced phototoxicity (Kuimova 2012).

5.3.2 Excitation and Emission

As mentioned earlier, different probes exhibit variations in their spectroscopic properties, particularly in terms of their excitation and emission spectra. Solvatochromic probes for measuring lipid packing in membranes give opportunity to choose depending on the excitation: Laurdan excites at around 340 nm in the UV spectrum, di-4-ANEPPDHQ excites at 488 nm and Nile Red based probes at 514 nm. This can be important when choosing a partner for colocalisation microscopy or availability of particular microscope setup. Also, one should consider that UV laser is cytotoxic and generated reactive oxygen species (Ramakrishnan, Maclean et al. 2016) therefore is not optimal for the live-cell imaging. It should be noted that because the dyes essentially have two emission spectra, each (and these are often broad) probes such as Laurdan and di-4-ANEPPDHQ can be difficult to multiplex with other fluorophores (such as fluorescent proteins for example). This is because they themselves occupy a large proportion of the visible spectrum. In general, multiplexing is done using the excitation spectrum instead. For example, di-4-ANQPPDHQ can be imaged with 488 nm excitation and simultaneous 560 and 610 nm detection. Then sequentially, mCherry can be imaged with 561 nm excitation and 610 nm detection.

5.3.3 Imaging Hardware

In general, there are two possible readouts when imaging environmentally sensitive dyes. These are spectral readouts and fluorescence lifetime readouts. For dyes that change their emission spectra (like Laurdan and di-4-ANEPPDHQ), spectral imaging is usually faster and simpler. This kind of imaging can be undertaken on any microscope system that can record in two spectral channels simultaneously. Confocal and multiphoton implementation is common because the use of point-detectors means the two channels are always registered. It is possible to use the dyes on wide-field systems too – either by using two cameras or an image splitter. Here, care must be taken that the two spectral channels are properly registered. This allows the dyes to be used with epi-fluorescence and

TIRF-based microscopes to study cell membranes (Owen, Oddos et al. 2010). Here though, it should be noted that the hard, high-refractive index glass coverslip might affect the spectral properties of the dye. Spectral-based readouts even allow the dyes to be used in the context of a cytometer or FACS machine (Miguel, Owen et al. 2011).

It should be remembered that the analysis is intrinsically ratiometric. This means that hardware that can influence the intensity in each channel should be treated with caution. On confocals, the voltage and gain applied to each PMT should not be varied between conditions in each experiment. Varying the voltage will naturally alter the ration of intensity. Similarly, with a 2-camera setup (e.g., for TIRF), the frame integration time and any gain should also be kept constant (although not necessarily equal). The use of a single camera and an image splitter avoids this issue, although does come at the expense of field-of-view.

For those dyes that do not change their emission spectra but instead change their lifetime, FLIM must be used. Here, there are no requirements for multi-channel imaging thus freeing detection pathways although the limitations on multiplexing noted earlier remain. For FLIM, TCSPC implemented on a confocal or multiphoton microscope is the most common method. In principle, the dyes can also be used in wide-field FLIM using a GOI or other method, or frequency domain FLIM. In these methods, however, one should be even more cautious of the glass coverslip affecting the fluorescence lifetimes, since lifetimes are intrinsically environmentally dependent, especially in the context of refractive index.

5.4 QUANTIFICATION

5.4.1 Ratiometric Imaging

For spectral readouts of Laurdan and di-4-ANEPPDHQ, we generally calculate a so-called generalised polarisation (GP) value. This calculation requires two intensity values as input. These can be averaged over a region of each image or can be from individual pixels. Using pixel-wise analysis allows the creation of a GP image. The GP is calculated as:

$$GP = \frac{I_B - I_R}{I_B + I_R}$$

Here, I_B represents the intensity in the blue channel (usually representing fluorescence emitted by the dye residing in an ordered, less fluid environment) and I_R represents the intensity collected in the red channel (usually originating from dye molecules residing in disordered, fluid areas of the membrane). Theoretically, the *GP* values can range from −1 (when all the intensity is in the red/disordered channel) to +1 (where all the intensity of from the blue/ordered channel). Note that the absolute intensity values and even their ratio have little meaning outside an experiment because factors like PMT gain, camera sensitivity curves, and filter transmission curves all affect the ratio. *GP*=0 therefore represents equal intensity in each channel, but not necessarily equal amounts of membrane order/disorder. As a general rule, the optimal setup is to have roughly equal intensity in each channel on average when imaging cells in the control condition. This gives the maximum dynamic range for *GP* values to vary up or down as treatments are performed.

5.4.2 Image Processing

Analysis software will calculate the *GP* values for each pixel in a 2-channel data set. These can then be pseudo-coloured to produce a *GP* image. Usually, high *GP* (high membrane order) will be coloured red and low *GP* (low membrane order) coloured blue. For example, the plasma membrane, which is rich in cholesterol, will generally be seen to have higher membrane order than intracellular membranes, which have a relatively low abundance of cholesterol. Despite the utility of this kind of visualisation, interpreting the images can be difficult. To help this, *GP* images are often merged with intensity image and displayed in a HSB colour space (this is also true of fluorescence lifetime images where the lifetime is colour coded). This means that the merged image displays two pieces of information – the membrane order (*GP* value or lifetime) and structural information (fluorescence intensity).

A final level of analysis is quantification. This involves extracting the *GP* values from at image, and for example, plotting a histogram of the *GP* values. This is often done by selecting a region-of-interest that the researcher wishes to study (e.g., the plasma membrane). This can be done manually or in an automated fashion (Panconi, Makarova et al. 2023). When performing this type of analysis, it is often useful to threshold the raw images – that is, disregard any pixels that have a low intensity in either channel. This is because low-intensity pixels have a low signal-to-noise ratio and this problem is exacerbated when taking a ratio. If the final *GP* image is noisy, it is possible to apply a Gaussian or other smoothing filter to the data to remove noise at the expense of spatial resolution.

5.5 POTENTIAL ARTEFACTS

5.5.1 Impact on Membrane Properties

Human cell membrane composition lies very close to the critical point of a ternary (unsaturated lipid, saturated lipids, cholesterol) phase diagram. The exact position of the composition relative to the critical point determines whether co-existing ordered and disordered phases will exist in the bilayer. Because of this proximity, the properties of the bilayer are very sensitive to changes in the composition. Most of the environmentally sensitive dyes that are used to study membranes are lipid-like in their structure. This means they usually have a polar head group, acyl tails and will insert themselves into the cell membrane. By doing this, they change the composition of the bilayer. If less dye is used, this effect will be small but will impact the signal-to-noise ratio that is achievable in an experiment. If more dye is used, the distortion to the membrane composition can be considerable.

It can be very difficult to know how dye molecules are affecting the biophysical properties of lipid bilayers. One way is to titrate the dye and measure various lipid properties at different concentrations. Another is to use atomistic scale molecular dynamics (MD) simulations. This has been done for PRODAN (a precursor of LAurdan), Laurdan and di-4-ANEPPDHQ (Suhaj, Le Marois et al. 2018; Suhaj, Gowland et al. 2020). The results show that the influence is complex with the dyes affecting *GP* values as well as other biophysical properties in different ways, depending on the underlying composition of the bilayer. Results from experiments using environmentally sensitive dyes should therefore always be interpreted with this in mind.

5.5.2 Image Analysis Artefacts

Besides the effect of the probes on the biophysical properties of membranes, there are a number of other potential artefacts that can arise. These are:

Photobleaching: Prolonged exposure to light can cause photobleaching, leading to a decrease in fluorescence intensity over time. This artefact can distort the measurements of membrane properties and hinder accurate analysis because it can change the ratio of fluorescence in each channel. Minimising the exposure time and optimising the imaging conditions can help mitigate this issue. Since the analysis involves calculating ratios, one also needs to be careful that image pixels are never saturated as this will distort the calculated *GP* values.

Uneven dye distribution: In some instances, environmentally sensitive dyes may exhibit uneven distribution on the cell membrane. This can occur due to variations in dye uptake or localisation within the cell. Uneven dye distribution can introduce artefacts, such as variations in fluorescence intensity across the membrane or false identification of membrane boundaries. Careful examination of multiple images, staining controls, and appropriate normalisation techniques can help compensate for this artefact. In general, it can be difficult to quantitatively compare membrane areas that have dramatically different intensities, e.g., because the contribution of background will change. In general, image analysis of environmentally sensitive dye images can be affected by background. Background can arise from various sources, including autofluorescence, impurities, and equipment limitations. This noise can hinder accurate detection and segmentation of cell membranes, leading to erroneous measurements.

5.6 BIOLOGICAL EXAMPLES

5.6.1 Immune Cells

Membrane lipid order has been studied in a wide variety of biological contexts. An example of the use of environmentally sensitive dyes like Laurdan and di-4-ANEPPDHQ is to study immune cells. In macrophages, Laurdan was used together with multiphoton microscopy to show that high membrane order resides at the phagocytic cup, an area of the membrane with dense cortical actin (Magenau, Benzing et al. 2011). In CD4+ Helper T cells, Laurdan was used to detect a ring of high membrane order at the immunological synapse – the interface between the T cell and the antigen presenting cell which is responsible for signal transduction and the regulation of T cell activation (Gaus, Chklovskaia et al. 2005). Using drugs like 7-ketocholesterol which disrupt membrane order, it was possible to show that this highly ordered region was crucial for the effective activation of these immune cells (Rentero, Zech et al. 2008). Di-4-ANEPPDHQ, which can be easily implemented in camera-based microscopes, was used to image high membrane order at the T cell synapse periphery using a 2-channel TIRF setup (Owen, Oddos et al. 2010). The synapse periphery is also dense with cortical actin, possibly indicating a link between high membrane order and the cytoskeleton. In the same study, sub-synaptic T cell vesicles were observed to have diverse levels of membrane order, including high- and low-order populations. In follow-up work, it was shown that these different populations of vesicles, defined by their membrane order move differently and carry different protein cargos (Ashdown, Williamson et al. 2018). In fact, T cells are notable for having a number of membrane proximal signalling proteins dependent on high order membrane domains for their localisation. Many of these have been studied by super-resolution microscopy including the scaffolding protein Linker for Activation of T cells (LAT; Williamson, Owen et al. 2011) and the Src-family kinase Lck (Rossy, Owen et al. 2013). Finally, di-4-ANEPPDHQ can also be implemented in a multi-channel flow cytometer. This gives the possibility to examine membrane lipid order on the population level across potentially millions of cells. It was shown that primary human T cells have a diverse range of membrane order that correlates with their function, possibly helping to define T cell subsets. High membrane cholesterol in T cells that are the corresponding high membrane order has been linked with autoimmune diseases including systemic lupus erythematosus (Miguel, Owen et al. 2011).

Membrane viscosity is extremely critical when it comes to diffusion rates of various membrane components. The frequencies of the protein–protein encounters are particularly relevant to cellular signalling and other cellular processes such as mitochondrial respiration (Budin, de Rond et al. 2018). Fluorescent molecular motors enabled the precise imaging of viscosity at the level of individual cell organelles, offering spatially resolved quantitative information, along with real-time dynamic viscosity measurements. Using this method, it has been shown that neurons after exposure to common neurodegenerative stimuli, such as excitotoxicity and oxidative stress, decrease their membrane viscosity which affects the mobility of proteins in the plasma membrane of these cells. Moreover, the membrane fluidification induced by oxidants was effectively countered by the H3 peptide, a wide-spectrum neuroprotectant. These findings provide insights into molecular mobility in neuronal membranes and the importance of preserving membrane stability for neuroprotection in brain disorders (Kubánková, Summers et al. 2019).

Most of the works with fluorogenic molecules are performed either on model membranes or in vitro on individual cells and it is challenging to address and measure the same parameters in vivo. However, for some probes, it has been done and the results replicate findings in vitro.

For example, a molecular rotor was used to measure tumour membrane viscosity in vivo (Shimolina, Izquierdo et al. 2017), this work proposed to use membrane viscosity as a signature of malignant state of the cells and this methodology offers a novel approach for the in vivo monitoring of tumour viscosity, presenting potential applications in diagnostics, treatment progress monitoring, and therapeutic interventions.

Using the membrane tension probe Flipper-TR, it has been shown that in mammalian cells peroxidation of lipids increases plasma membrane tension and leads to the activation of mechanosensitive cation channels such as Piezo. These findings demonstrate key roles of mechanosensing channels in the execution of ferroptosis (Hirata, Cai et al. 2023).

Osmotic adaptation of cells requires mechanosensing of membrane tension, and using Flipper-TR, it has been shown that cells can regulate their volume in response to osmotic shock and primarily regulate this process by mTORC2. Specifically, it has been shown that the coupling between membrane volume and tension changes during osmotic adaptation. And this coupling is actively regulated by cellular components, including the cytoskeleton, ion transporters, and mTOR pathways (Roffay, Molinard et al. 2021).

REFERENCES

Amaro, M., F. Reina, M. Hof, C. Eggeling and E. Sezgin (2017). "Laurdan and Di-4-ANEPPDHQ probe different properties of the membrane." *Journal of Physics D: Applied Physics* **50**(13): 134004.

Andersen, O. S. and I. Roger E. Koeppe (2007). "Bilayer thickness and membrane protein function: an energetic perspective." *Annual Review of Biophysics and Biomolecular Structure* **36**(1): 107–130.

Ashdown, G. W., D. J. Williamson, G. H. M. Soh, N. Day, G. L. Burn and D. M. Owen (2018). "Membrane lipid order of sub-synaptic T cell vesicles correlates with their dynamics and function." *Traffic* **19**(1): 29–35.

Atilla-Gokcumen, G. E., E. Muro, J. Relat-Goberna, S. Sasse, A. Bedigian, Margaret L. Coughlin, S. Garcia-Manyes and Ulrike S. Eggert (2014). "Dividing cells regulate their lipid composition and localization." *Cell* **156**(3): 428–439.

Bagatolli, L. A. (2006). "To see or not to see: lateral organization of biological membranes and fluorescence microscopy." *Biochimica et Biophysica Acta (BBA) - Biomembranes* **1758**(10): 1541–1556.

Barceló-Coblijn, G., M. L. Martin, R. F. M. de Almeida, M. A. Noguera-Salvà, A. Marcilla-Etxenike, F. Guardiola-Serrano, A. Lüth, B. Kleuser, J. E. Halver and P. V. Escribá (2011). "Sphingomyelin and sphingomyelin synthase (SMS) in the malignant transformation of glioma cells and in 2-hydroxyoleic acid therapy." *Proceedings of the National Academy of Sciences* **108**(49): 19569–19574.

Budin, I., T. de Rond, Y. Chen, L. J. G. Chan, C. J. Petzold and J. D. Keasling (2018). "Viscous control of cellular respiration by membrane lipid composition." *Science* **362**(6419): 1186–1189.

Colom, A., E. Derivery, S. Soleimanpour, C. Tomba, M. D. Molin, N. Sakai, M. González-Gaitán, S. Matile and A. Roux (2018). "A fluorescent membrane tension probe." *Nature Chemistry* **10**(11): 1118–1125.

Danylchuk, D. I., P.-H. Jouard and A. S. Klymchenko (2021). "Targeted solvatochromic fluorescent probes for imaging lipid order in organelles under oxidative and mechanical stress." *Journal of the American Chemical Society* **143**(2): 912–924.

Desmond, E. and S. Gribaldo (2009). "Phylogenomics of sterol synthesis: insights into the origin, evolution, and diversity of a key eukaryotic feature." *Genome Biology and Evolution* **1**: 364–381.

Gaus, K., E. Chklovskaia, B. Fazekas de St Groth, W. Jessup and T. Harder (2005). "Condensation of the plasma membrane at the site of T lymphocyte activation." *Journal of Cell Biology* **171**(1): 121–131.

Harayama, T. and H. Riezman (2018). "Understanding the diversity of membrane lipid composition." *Nature Reviews Molecular Cell Biology* **19**(5): 281–296.

Hirata, Y., R. Cai, A. Volchuk, B. E. Steinberg, Y. Saito, A. Matsuzawa, S. Grinstein and S. A. Freeman (2023). "Lipid peroxidation increases membrane tension, Piezo1 gating, and cation permeability to execute ferroptosis." *Current Biology* **33**(7): 1282–1294.e1285.

Jin, L., A. C. Millard, J. P. Wuskell, X. Dong, D. Wu, H. A. Clark and L. M. Loew (2006). "Characterization and application of a new optical probe for membrane lipid domains." *Biophysical Journal* **90**(7): 2563–2575.

Krogh, A., B. Larsson, G. von Heijne and E. L. L. Sonnhammer (2001). "Predicting transmembrane protein topology with a hidden markov model: application to complete genomes11Edited by F. Cohen." *Journal of Molecular Biology* **305**(3): 567–580.

Kubánková, M., P. A. Summers, I. López-Duarte, D. Kiryushko and M. K. Kuimova (2019). "Microscopic viscosity of neuronal plasma membranes measured using fluorescent molecular rotors: effects of oxidative stress and neuroprotection." *ACS Applied Materials & Interfaces* **11**(40): 36307–36315.

Kuimova, M. K. (2012). "Mapping viscosity in cells using molecular rotors." *Physical Chemistry Chemical Physics* **14**(37): 12671–12686.

Kuimova, M. K., G. Yahioglu, J. A. Levitt and K. Suhling (2008). "Molecular rotor measures viscosity of live cells via fluorescence lifetime imaging." *Journal of the American Chemical Society* **130**(21): 6672–6673.

Kwiatek, J. M., D. M. Owen, A. Abu-Siniyeh, P. Yan, L. M. Loew and K. Gaus (2013). "Characterization of a new series of fluorescent probes for imaging membrane order." *PLOS One* **8**(2): e52960.

Leventan, I. and E. Lyman (2023). "Regulation of membrane protein structure and function by their lipid nano-environment." *Nature Reviews Molecular Cell Biology* **24**(2): 107–122.

López-Duarte, I., T. T. Vu, M. A. Izquierdo, J. A. Bull and M. K. Kuimova (2014). "A molecular rotor for measuring viscosity in plasma membranes of live cells." *Chemical Communications* **50**(40): 5282–5284.

Magenau, A., C. Benzing, N. Proschogo, A. S. Don, L. Hejazi, D. Karunakaran, W. Jessup and K. Gaus (2011). "Phagocytosis of IgG-coated polystyrene beads by macrophages induces and requires high membrane order." *Traffic* **12**(12): 1730–1743.

Miguel, L., D. M. Owen, C. Lim, C. Liebig, J. Evans, A. I. Magee and E. C. Jury (2011). "Primary human CD4+ T cells have diverse levels of membrane lipid order that correlate with their function." *Journal of Immunology* **186**(6): 3505–3516.

Owen, D. M., P. M. P. Lanigan, C. Dunsby, I. Munro, D. Grant, M. A. A. Neil, P. M. W. French and A. I. Magee (2006). "Fluorescence lifetime imaging provides enhanced contrast when imaging the phase-sensitive dye di-4-ANEPPDHQ in model membranes and live cells." *Biophysical Journal* **90**(11): L80–L82.

Owen, D. M., A. Magenau, A. Majumdar and K. Gaus (2010). "Imaging membrane lipid order in whole, living vertebrate organisms." *Biophysical Journal* **99**(1): L7–L9.

Owen, D. M., S. Oddos, S. Kumar, D. M. Davis, M. A. Neil, P. M. French, M. L. Dustin, A. I. Magee and M. Cebecauer (2010). "High plasma membrane lipid order imaged at the immunological synapse periphery in live T cells." *Molecular Membrane Biology* **27**(4–6): 178–189.

Owen, D. M., C. Rentero, A. Magenau, A. Abu-Siniyeh and K. Gaus (2012). "Quantitative imaging of membrane lipid order in cells and organisms." *Nature Protocols* **7**(1): 24–35.

Owen, D. M., D. J. Williamson, A. Magenau and K. Gaus (2012). "Sub-resolution lipid domains exist in the plasma membrane and regulate protein diffusion and distribution." *Nature Communications* **3**(1): 1256.

Panconi, L., M. Makarova, E. R. Lambert, R. C. May and D. M. Owen (2023). "Topology-based fluorescence image analysis for automated cell identification and segmentation." *Journal of Biophotonics* **16**(3): e202200199.

Parasassi, T., G. De Stasio, G. Ravagnan, R. M. Rusch and E. Gratton (1991). "Quantitation of lipid phases in phospholipid vesicles by the generalized polarization of Laurdan fluorescence." *Biophysical Journal* **60**(1): 179–189.

Pliotas, C., A. C. E. Dahl, T. Rasmussen, K. R. Mahendran, T. K. Smith, P. Marius, J. Gault, T. Banda, A. Rasmussen, S. Miller, C. V. Robinson, H. Bayley, M. S. P. Sansom, I. R. Booth and J. H. Naismith (2015). "The role of lipids in mechanosensation." *Nature Structural & Molecular Biology* **22**(12): 991–998.

Preta, G. (2020). "New insights into targeting membrane lipids for cancer therapy." *Frontiers in Cell and Developmental Biology* **8**: 571237.

Ramakrishnan, P., M. Maclean, S. J. MacGregor, J. G. Anderson and M. H. Grant (2016). "Cytotoxic responses to 405 nm light exposure in mammalian and bacterial cells: involvement of reactive oxygen species." *Toxicology in Vitro* **33**: 54–62.

Rentero, C., T. Zech, C. M. Quinn, K. Engelhardt, D. Williamson, T. Grewal, W. Jessup, T. Harder and K. Gaus (2008). "Functional implications of plasma membrane condensation for T cell activation." *PLoS One* **3**(5): e2262.

Roffay, C., G. Molinard, K. Kim, M. Urbanska, V. Andrade, V. Barbarasa, P. Nowak, V. Mercier, J. García-Calvo, S. Matile, R. Loewith, A. Echard, J. Guck, M. Lenz and A. Roux (2021). "Passive coupling of membrane tension and cell volume during active response of cells to osmosis." *Proceedings of the National Academy of Sciences of the United States of America* **118**(47): 1–12.

Rossy, J., D. M. Owen, D. J. Williamson, Z. Yang and K. Gaus (2013). "Conformational states of the kinase Lck regulate clustering in early T cell signaling." *Nature Immunology* **14**(1): 82–89.

Shimolina, L. E., M. A. Izquierdo, I. López-Duarte, J. A. Bull, M. V. Shirmanova, L. G. Klapshina, E. V. Zagaynova and M. K. Kuimova (2017). "Imaging tumor microscopic viscosity in vivo using molecular rotors." *Scientific Reports* **7**(1): 41097.

Simons, K. and E. Ikonen (1997). "Functional rafts in cell membranes." *Nature* **387**(6633): 569–572.

Singer, S. J. and G. L. Nicolson (1972). "The fluid mosaic model of the structure of cell membranes." *Science* **175**(4023): 720–731.

Suhaj, A., D. Gowland, N. Bonini, D. M. Owen and C. D. Lorenz (2020). "Laurdan and Di-4-ANEPPDHQ influence the properties of lipid membranes: a classical molecular dynamics and fluorescence study." *The Journal of Physical Chemistry B* **124**(50): 11419–11430.

Suhaj, A., A. Le Marois, D. J. Williamson, K. Suhling, C. D. Lorenz and D. M. Owen (2018). "PRODAN differentially influences its local environment." *Physical Chemistry Chemical Physics* **20**(23): 16060–16066.

Sunshine, H. and M. L. Iruela-Arispe (2017). "Membrane lipids and cell signaling." *Current Opinion in Lipidology* **28**(5): 408–413.

Williamson, D. J., D. M. Owen, J. Rossy, A. Magenau, M. Wehrmann, J. J. Gooding and K. Gaus (2011). "Pre-existing clusters of the adaptor Lat do not participate in early T cell signaling events." *Nature Immunology* **12**(7): 655–662.

CHAPTER 6

Conclusions and Future Perspectives

6.1 SUMMARY OF CONCLUSIONS

6.1.1 Single-Molecule Localisation Microscopy

Single-molecule localisation microscopy (SMLM) has emerged as a powerful technique for studying the nanoscale distribution of membrane properties. Traditional optical microscopy is limited by the diffraction barrier, preventing the direct observation of structures smaller than half the wavelength of light. However, SMLM techniques such as photoactivated localisation microscopy (PALM) and stochastic optical reconstruction microscopy (STORM) have overcome this limitation by exploiting the stochastic blinking behaviour of individual fluorophores. SMLM enables the visualisation of individual fluorophores with nanometre precision, providing unprecedented insights into the spatial organisation and dynamics of membrane properties. By labelling specific membrane components or using genetically encoded tags, researchers can probe the distribution of proteins, lipids, and other biomolecules on the cell membrane.

One of the key advantages of SMLM is its ability to uncover nanoscale heterogeneities in membrane organisation. By analysing the precise localisation of individual fluorophores, SMLM reveals intricate patterns and spatial variations that were previously inaccessible. For example, studies have shown the presence of nanoscale protein clusters or microdomains on the cell membrane, such as lipid rafts or signalling platforms. SMLM has

DOI: 10.1201/9781003273745-6

provided direct evidence for the existence of these structures and allowed for their characterisation in terms of size, density, and composition. SMLM can also be combined with other techniques to gain further insights into membrane properties, for example, by integrating single-molecule imaging with fluorescence resonance energy transfer (FRET).

SMLM has been applied to investigate membrane organisation in immune cells, neuronal synapses, and subcellular compartments. By studying the nanoscale distribution of membrane properties in these specific contexts, researchers can gain insights into their functional implications. For instance, SMLM has revealed the clustering and organisation of immune signalling molecules in immune cells, shedding light on their role in immune response regulation.

6.1.2 Imaging Membrane Protein Dynamics: FCS, SPT, and FRAP

Fluorescence recovery after photobleaching (FRAP) is a widely used technique for studying the diffusion of membrane proteins. This non-invasive method provides valuable insights into the mobility and dynamics of proteins within the cell membrane. FRAP involves selectively photobleaching a region of interest within a fluorescently labelled membrane protein and then monitoring the recovery of fluorescence over time as unbleached molecules move into the bleached area. By analysing the kinetics of fluorescence recovery, researchers can extract information about the diffusion properties of the protein of interest such as the diffusion coefficient, mobile fraction, and residence time of proteins within the membrane. This information provides insights into the interactions, clustering, and compartmentalisation of membrane proteins, as well as their association with lipid rafts or other membrane microdomains. FRAP can also be used to investigate the influence of various factors on protein diffusion. For example, researchers can examine how changes in temperature or the presence of specific ligands or signalling molecules affect the mobility of membrane proteins. Additionally, FRAP can reveal alterations in protein diffusion due to post-translational modifications or changes in protein–protein interactions. Despite this, one of the key disadvantages of FRAP is its inability to determine the mobility of membrane proteins at the individual molecule level.

Single-particle tracking (SPT) is a powerful technique for studying the diffusion of membrane proteins at the level of individual molecules. By tracking the movement of fluorescently labelled proteins in real-time, SPT provides detailed information about their mobility, interactions, and spatial organisation within the cell membrane. SPT involves the labelling of membrane proteins with fluorophores and the use of high-resolution microscopy to track their trajectories over time. By analysing the displacement, speed, and directionality of individual particles, researchers can characterise the diffusion properties of membrane proteins.

One of the key advantages of SPT is its ability to capture the heterogeneous nature of protein diffusion. It reveals variations in diffusion coefficients among different proteins or within distinct regions of the cell membrane. This information can shed light on the presence of barriers, confinement, or compartmentalisation of proteins within the membrane. SPT also allows for the investigation of transient interactions and binding events involving membrane proteins. By analysing the dwell times and spatial proximity of particles, researchers can gain insights into protein–protein interactions, protein–lipid interactions, or the formation of protein complexes. This information is crucial for understanding the functional implications of protein diffusion and its role in cellular processes.

Fluorescence Correlation Spectroscopy (FCS) can also be employed to study the diffusion of membrane proteins. FCS relies on the principles of correlation analysis. By labelling membrane proteins of interest with fluorescent markers, their movement can be tracked and quantified. FCS measures the fluctuation in fluorescence intensity caused by the diffusion of labelled molecules within a small observation volume. This allows the determination of parameters such as diffusion coefficients, concentrations, and molecular interactions.

To study membrane protein diffusion using FCS, a small region of interest on the cell membrane is illuminated with a laser beam, and the resulting fluorescence is recorded over time. The recorded data is then subjected to autocorrelation analysis, which calculates the correlation between fluorescence intensity fluctuations at different time intervals. From the autocorrelation function, various parameters related to diffusion can be extracted, such as the diffusion coefficient.

6.1.3 FRET

FRET has revolutionised the study of molecular interactions and spatial organisation within biological systems. It is particularly well-suited for investigating the clustering of membrane proteins, which play crucial roles in various cellular processes such as signal transduction, cell adhesion, and membrane trafficking. This is especially true when the proteins interact as small oligomers, such as dimers, which can be difficult to study using SMLM.

FRET is based on the transfer of energy between two fluorophores: a donor and an acceptor. The donor fluorophore is excited by a light source, and the energy can be non-radiatively transferred to the nearby acceptor fluorophore if certain conditions are met, such as sufficient spectral overlap and proximity between the two molecules. By labelling the membrane proteins of interest with appropriate fluorophores, FRET can be employed to investigate their spatial distribution and interactions. By analysing the changes in FRET efficiency, researchers can infer the presence and extent of protein clustering. Higher FRET efficiencies indicate closer proximity between the labelled proteins, suggesting the formation of protein clusters. FRET can also be used to study conformational changes within the same molecule.

Several FRET-based techniques have been developed to investigate membrane protein clustering. For instance, fluorescence lifetime imaging microscopy (FLIM) measures the fluorescence decay time of the donor fluorophore. In protein clusters, FRET leads to a reduction in donor lifetime. By quantifying these changes, FLIM-FRET can provide spatial information about the nanoscale distribution of proteins. Single-molecule FRET (smFRET) and super-resolution microscopy can also provide detailed insights into protein clustering at the nanoscale level. These methods enable the visualisation and quantification of individual protein interactions within clusters, allowing for a more precise understanding of their organisation and function.

6.1.4 Environmentally Sensitive Dyes

Environmentally sensitive dyes have emerged as valuable tools for studying the biophysical properties of cell membranes. These dyes, such as Laurdan and di-4-ANEPPDHQ, are sensitive to the physical and chemical properties of the surrounding lipid environment, allowing researchers to probe aspects like membrane fluidity, polarity, and organisation.

Laurdan and di-4-ANEPPDHQ are two commonly used environmentally sensitive dyes. These dyes possess unique spectral properties that undergo changes depending on the physicochemical properties of the lipid bilayer in which they are embedded. Both exhibit a shift in its emission spectrum depending on the polarity of its surrounding environment. In non-polar regions of the lipid bilayer, such as the hydrophobic core, Laurdan emits blue light. In Polar Regions, such as the water–lipid interface, it emits green light. By measuring the emission intensity at these two wavelengths, researchers can calculate the generalised polarisation

(GP) value, which reflects the relative polarity of the lipid environment. The GP value provides information about the fluidity and organisation of the membrane. Higher GP values indicate a more ordered and rigid membrane, while lower values indicate a more fluid and disordered state. By comparing the GP values of different cellular regions or experimental conditions, researchers can assess changes in membrane fluidity and organisation, providing insights into processes such as lipid rafts formation, membrane phase transitions, and lipid–protein interactions.

Additionally, these environmentally sensitive dyes can be used in various experimental setups, including live-cell imaging, flow cytometry, and spectroscopy. This versatility allows for the investigation of membrane properties in different cellular contexts and experimental conditions.

6.2 FUTURE PERSPECTIVES

Fluorescence microscopes generally require a trade-off between spatial and temporal resolution, imaging volume, live-cell compatibility, and even cost and complexity. Up and coming developments are likely to either provide new instruments which balance these factors, or push the binderies of one or more of these properties. One example is the ongoing development of light-sheet microscopy. Like confocal microscopes, these are specialised to collect three-dimensional data; however, as they do not require point-scanning, they are dramatically faster and deliver lower radiation doses to the cells, so live cells can be imaged for longer. These properties mean these microscopes will likely challenge the dominance of confocal microscopes in research labs and imaging facilities (Chen, Chang et al. 2022). For this to happen, however, there will likely need to be progress in simplifying the usability of the systems which often require more intricate sample preparation, microscope alignment, and image post-processing.

An area of rapid development is pushing the spatial resolution to the nanometre level and beyond. While this often comes at significant extra cost in terms of complexity, cost, and experimental throughput, techniques such as MINFLUX now allow proteins to be localised to precisions which border on the sizes of the molecules themselves (Balzarotti, Eilers et al. 2017). Methods such as ONE, which combined expansion microscopy with advanced analysis, can achieve similar performance (Shaib et al. 2022). Other developments include enhanced multiplexing, which allows many protein species to be imaged in order to study their interactions (Unterauer et al. 2023).

Another area of development is correlated light and electron microscopy (CLEM; de Boer, Hoogenboom et al. 2015). These methods seek to combine the advantages of fluorescence (molecular specificity due to the labelling with fluorophores) with electron microscopy (structural details at the highest possible resolution). CLEM is gaining popularity despite dispensing with one of the major advantages of fluorescence imaging, namely live-cell compatibility and dynamic imaging. For example, it would be possible to use EM to provide nanometre resolution information on the positions of membranes and their thickness or curvature and correlate that information with the location of specific proteins of interest, such as signalling proteins or adhesion molecules.

Despite the insights provided by advanced imaging techniques, there is still much we do not know about membrane composition, organisation structure, and function. In general, we still do not fully understand how membrane biophysical properties regulate protein–protein interactions and hence signalling at the membrane. How does the cell regulate membrane composition and monitor membrane phase behaviour in order to control protein interactions? There are also questions about how cells regulate their lipid metabolic machinery to generate the required lipidomes and how they are able to alter these in order to adapt to changing environmental conditions. Here, there may be a role for specific, understudied lipid molecules (Smith, Owen et al. 2021).

REFERENCES

Balzarotti, F., Y. Eilers, K. C. Gwosch, A. H. Gynnå, V. Westphal, F. D. Stefani, J. Elf and S. W. Hell (2017). "Nanometer resolution imaging and tracking of fluorescent molecules with minimal photon fluxes." *Science* 355(6325): 606–612.

Chen, B., B.-J. Chang, P. Roudot, F. Zhou, E. Sapoznik, M. Marlar-Pavey, J. B. Hayes, P. T. Brown, C.-W. Zeng, T. Lambert, J. R. Friedman, C.-L. Zhang, D. T. Burnette, D. P. Shepherd, K. M. Dean and R. P. Fiolka (2022). "Resolution doubling in light-sheet microscopy via oblique plane structured illumination." *Nature Methods* 19(11): 1419–1426.

de Boer, P., J. P. Hoogenboom and B. N. G. Giepmans (2015). "Correlated light and electron microscopy: ultrastructure lights up!" *Nature Methods* 12(6): 503–513.

Shaib, A. H. et al. (2022). "Expansion microscopy at one nanometer resolution." *BioRxiv*. https://doi.org/10.1101/2022.08.03.502284.

Smith, P., D. M. Owen, C. D. Lorenz and M. Makarova (2021). "Asymmetric glycerophospholipids impart distinctive biophysical properties to lipid bilayers." *Biophysical Journal* 120(9): 1746–1754.

Unterauer, E. M. et al. (2023). "Spatial proteomics in neurons at single-protein resolution." *BioRxiv*. https://doi.org/10.1101/2023.05.17.541210.

Index